Ten Million Aliens

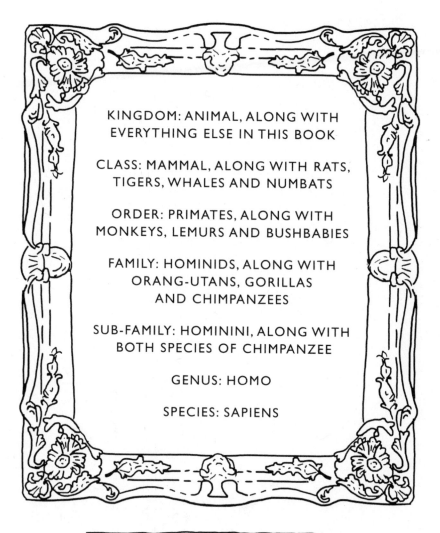

KINGDOM: ANIMAL, ALONG WITH
EVERYTHING ELSE IN THIS BOOK

CLASS: MAMMAL, ALONG WITH RATS,
TIGERS, WHALES AND NUMBATS

ORDER: PRIMATES, ALONG WITH
MONKEYS, LEMURS AND BUSHBABIES

FAMILY: HOMINIDS, ALONG WITH
ORANG-UTANS, GORILLAS
AND CHIMPANZEES

SUB-FAMILY: HOMININI, ALONG WITH
BOTH SPECIES OF CHIMPANZEE

GENUS: HOMO

SPECIES: SAPIENS

HOW TO CLASSIFY A HUMAN

Ten Million Aliens

A Journey Through the Entire Animal Kingdom

SIMON BARNES

MARBLE ARCH
PRESS

Marble Arch Press
1230 Avenue of the Americas
New York, NY 10020

First Marble Arch Press hardcover edition February 2015

Marble Arch Press is a publishing collaboration between Short Books, UK and Atria Books, US.

Marble Arch Press and colophon are trademarks of Short Books.

For information about special discounts for bulk purchases, please contact Simon & Schuster Special Sales at 1-866-506-1949 or business@simonandschuster.com

Cover design and illustrations
by nathanburtondesign.com

Manufactured in the United States of America

10   9   8   7   6   5   4   3   2

Library of Congress Cataloging-in-Publication Data
has been applied for.

ISBN 978-1-4767-3035-6
ISBN 978-1-4767-3036-3 (ebook)

To the great teachers --

especially

*Mr Hendry – "Pete" – late of Emanuel School, who
taught me about words and books and Joyce,*

*Sir David Attenborough, who taught me about the wild
world and Darwin,*

and CLW who taught me practically everything

Contents

INVERTEBRATE CYCLE

VERTEBRATE CYCLE

Here Comes Everybody
James Joyce, Finnegans Wake

Endlessness

endless forms most beautiful and most wonderful have been, and are being, evolved. Final words of *The Origin of Species* by Charles Darwin. It is a thought that has had me enthralled all my life. We are not alone in the universe: the idea that launched a million works of science fiction. Fact is we are not alone on our own planet. Far from it. We could hardly be less alone. We are one of a crowd, part of a teeming throng. We are not alone even when we are alone: whether we are counting the great garden of bacteria in our guts – alien life forms that keep us alive – or the tiny arthropods called Demodex mites that live in the follicles of our eyelashes.

Because we are one of many. Life is not about the creation of a single perfect being. An ape is not a failed human: it is a perfectly valid and fully evolved creature in its own right. A monkey is not a failed ape, a lemur is not a failed monkey, a mouse is not a failed primate, a fish is not a failed mammal (and as I shall show you later, there is no such thing as a fish) and insects, nematode worms, corals and priapulids are not failed vertebrates. The meaning of life is life and the purpose of life is to become an ancestor. All forms of life are equally valid: the beautiful, the bizarre, the horrific, the obscure and the glorious.

We humans are different from the rest in some ways, but only in some ways. One of these ways is our need for a myth to get us through the night: a myth to carry us through the vast distances of interstellar space: a myth to transport us through the endless aeons of time in which life has been lived on earth, a myth to reconcile us to our true evolutionary position. Which is a cosmic afterthought.

We used to cherish the myth that we are made of quite different stuff from the animals: there are animals, and then there's us. Darwin exploded that one, of course. He showed us that we are all animals, but that is too difficult a truth for us to face in its rawness and reality. So we have created another myth. Benjamin Disraeli, speechifying about Darwin's horrifying truth, said: "The question is this: is man an ape or an angel? I, my lord, am on the side of the angels. I repudiate with indignation and abhorrence these new-fangled theories."

But evolution is a fact and we humans – let's dispense with Disraeli's "man" nonsense; we're all in it together, men, women and children – needed to come to terms with our apeness, our primateness, our mammalness, our vertebrateness, our animalness. So we came up with perfectibility: the idea that evolution had a goal, that goal was to make a perfect creature, and that perfect creature is lucky old us. The famous image of evolution – monkey, ape, hunched proto-hominid, fully evolved and upright modern man – encapsulates the myth as vividly as a cross, a crescent and a seated Buddha encapsulate the great world religions. The whole process of the Animal Kingdom, starting with unicellular blobs and passing through insects, "fish", amphibians, reptiles and birds, culminates in mammals, and mammals carry us through primitive egg-layers and marsupials, to creatures of ever-greater magnificence and complexity, to the primates and then the apes, until the ladder finally ascends to wonderful, glorious, magical us.

Which is great. Except of course that it doesn't.

The mite that lives in the follicles of your eyelashes is as fully, as exquisitely, as perfectly evolved as you are. And on that thought, I shall set out to describe the endless forms

I. HUMAN

of the Animal Kingdom,* to encounter the ten million alien species with which we share our planet. To do so righteously, I must write a book that has no beginning and no end, but like James Joyce's *Finnegans Wake*, simply continues.

Ten million aliens, then. Or is there really only one? Perhaps the only alien species on the planet is us. Certainly we are alienated from the rest of creation: so much so that we have become tourists on our own planet. What all tourists need is a travel guide: so here you are. Look on this book as the Rough Guide to Real Life: the Lonely Planet Guide to the Lonely Planet.

---

* In this book I'm modestly restricting myself to the Animal Kingdom. In Britain four other kingdoms are traditionally recognised: Plantae, Fungi, Protoctista and Prokaryota/Monera. In the United States they prefer six: Animalia, Plantae, Fungi, Protista, Archaea and Bacteria.

Sex and the single slug

This is the ideal point in the book for a quick lesson in basic taxonomy, so let's talk about sex instead. I may not be able to teach you how to love slugs, but I can certainly teach you how slugs love. Slugs are invertebrates – molluscs, since you ask – and as such, less than the trash beneath a vertebrate's feet. But they live lives of sensuous and sometimes violent passions and go in for the most bewildering and perhaps even enviable gymnastics.

We are brought up to despise most invertebrates, especially the slimy ones. I quote an exchange in Australia's federal parliament in 2006. Julia Gillard, a Labour frontbencher: "Mr Speaker, I move that that snivelling grub over there be no further heard." Speaker: "The manager of the opposition will withdraw that." Gillard: "If I have offended grubs I withdraw unconditionally."[*]

But we should warm to slugs if only for their sex lives: creatures whose antics outdo anything thought up by Messalina, Zeus and all those carved and writhing figures on the Konark Temple. As gardeners wage war on slugs with beer and eggshells, so slugs pursue their exotic and passionate lives. There are about 5,000 species of slugs in the world, and 32 of them in Britain, where the cold limits the things a slug can get up to. Slugs are related to snails, and are not unlike shell-less snails, except that to be confusing – and life and biodiversity are confusing almost by design[†]

---

[*] Extracted from *Extracts from the Red Notebooks* by my old friend Matthew Engel.
[†] Except that there is no design.

– there are three species of shelled slugs in Britain. Being molluscs, slugs are related to giant squids – but we'll save them for later.

Slugs have two pairs of tentacles; the front ones sense light and the back ones sense smell. These are retractable, and they can be regrown. And yes, they do slime. Two sorts of slime: watery stuff, and thick, sticky stuff. They get about by gliding gracefully along this self-created carpet. It's hard for humans to get excited about mucus – though it is life and death to slugs – so let's move on to sex.

Slugs are hermaphrodites. Both halves of a pair have penises, both halves present sperm to the partner, and both halves go off and lay eggs. Slugs have the best of both worlds. But they are not just wham-bammers. They believe in courtship. Perhaps, being female as well as male, they are devoted, to the point of mania, to the concept of foreplay. It can go on for hours, as they circle, nibble and lunge at each other. Sometimes they will savour each other's mucus, perhaps to get genetic information, perhaps as a light sustaining snack. Anointed in mucus, they engage in a slimy and sensual ballet. Some species will do this suspended from long ropes of mucus: acrobatic, gravity-defying, and no doubt as thrilling as doing it on a trapeze. The pace is slow, the rhythm sensuous, as if each nuance is relished. And some species have the most colossal penises: half as long as their own bodies. Some slugs have copulatory rituals in which the pair dance about each other, each partner waving a giant penis overhead.

The act continues with a mutual entering and a prolonged and slimy embrace. But then, how to break it off? The phrase, I fear, is no metaphor. With some species, a long and corkscrew-shaped penis doesn't always withdraw too easily. In these circumstances – gentlemen are invited to cross their legs at this point – one slug will chew off the penis of the other. Sometimes both slugs will perform this feat. It is called apophallation. The slug, hermaphrodite no longer, goes away. Alas, he can't grow another penis. So she carries on as a female forever after.

A backbone isn't essential to an interesting life.

2, 8, 10, 12, 13, 18

I know we should be getting on with that lesson in taxonomy, but let's think about women's breasts instead. Turn to page three of the *Sun*, have a leaf through *Playboy* magazine, find pictures of naked women on the internet, look at *Le déjeuner sur l'herbe* or *The Birth of Venus*. Lots of breasts. Every two maintaining the most perfect paradox.

What are humans? Whatever else we are, we are a species of mammal, and as such we suckle our young. The female half of our species and the female half of the class of mammals all possess mammary glands – in one form or another. Humans, goats, sheep, horses, elephants and guinea pigs have two teats; dogs have eight or ten, rats have 12, pigs have 18, while the Virginia opossum has 13, one of the few mammals to have an odd number. The primitive egg-laying mammals, the monotremes, have no teats at all but they can still sweat milk. In other words, we're mammals. All of us. That simple fact was never explained to me in these straightforward and uncompromising terms. I was taught that we "come from" mammals, or "come from" the apes. But we *are* mammals: so let's deal with it. Some of the greatest art ever produced celebrates the mammalian defining characteristic of mammary glands, in suckling madonnas and in succulent nudes.

But let us also consider the contradiction. Humans are the only mammal that uses mammary glands for sexual display. There are plenty of theories about this: the most frequent is the notion that when humans started walking upright, the bottom became a lot less obvious. Female baboons signal sexual availability with a red flush to the buttocks: a trait which is noticeable, even spectacular, to passing humans. Non-female readers can guess the effect that rose-red buttocks have on a male baboon with male-to-male empathy. But with the bottom going out of fashion as a sexual signal, humans developed spectacular signalling breasts. They didn't need to flush on and off, because human females were and are always sexually available.

Physiologically, physiologically. Not morally, not morally at all. I hope that's obvious. Physiology is not the same as morality; physiology does not imply morality. We need to be clear about that before we move on any further. There is an awful lot of difference between I can and I must. Humans are capable of making personal and moral choices: if there is any discontinuity between humans and other animals, it is probably to be found here – though some primatologists suggest that morality has its origin in the responsibilities and obligations of social life and can be found in non-human societies.*

We have the technology to control our rates of reproduction and the ability to make the consequent choices. Women can make their own decisions about sexual and reproductive availability.†

Decisions about reproduction are crucial to the future of the human species and of the planet we live on: overpopulation is the biggest single problem we face.

We can make choices, then: but that doesn't stop us from being animals. Breasts and our preoccupation with them are

---

* See for example *The Bonobo and the Atheist* by Frans de Waal.
† In Stella Gibbons's novel *Cold Comfort Farm*, the heroine Flora gives advice to Meriam, the hired girl, when she asks: "'Who's to know what will happen to me when the sukebind is out in the hedges again and I feel so strange on the long summer evenings –'
'Nothing will happen to you if only you use your intelligence and see that it doesn't,' Flora said. [Flora then explains the precautionary arts.]
"'Tes wickedness! 'Tes flying in the face of nature!' she burst out fearfully at last.
'Nonsense,' said Flora. 'Nature is all very well in her place, but she must not be allowed to make things untidy.'"

at once an emblem of the human continuity with mammals and therefore with the rest of the Animal Kingdom – and at the same time, an emblem of human uniqueness. This book is mostly about continuities and connections, but I am not here to deny human uniqueness: that would be perverse. The point here is that, though we may be unique, we are certainly not separate.

Champagne lifestyle

I used to think they cracked open a bottle of champagne every time they discovered a new species. New to science! Glorious phrase, dizzying thought. Imagine finding a creature that nobody knew about. The honour and glory of it would last you throughout all eternity. Naturally you'd open the champagne.

But they don't. For a start, if they did so they'd be pissed all day. And the second thing is that there isn't really any "they". When I was young I sort of assumed that somewhere there must be a book – a nice big fat book – that contained all the species in the world. The Book of Life: nothing less. Occasionally, very occasionally, the Book would need a few solemn alterations and then a new version, bigger, deeper and truer than anything that had gone before, would be printed in its stead. But there isn't. There isn't even a definitive database. There isn't a definitive anything. There can't be.

Which is an even more dizzying thought. I'm a human being: that's my dog and that's a blackbird. Three species: what could be simpler? But it turns out that life isn't neat and tidy at all.

I once tried to establish a new species myself. The full story is told elsewhere.* It was a small brown bird. I and three others made an expedition into the Northwestern province of Zambia to try and make observations and recordings that would demonstrate unequivocally that

---

* *How to Be Wild*, published by Short Books in 2007.

this little bird was not a subspecies of the nedicky but a good species called Pearson's cisticola. Had we found the evidence,* one of us would have submitted a paper to the Zambian Ornithological Society. That, we hoped, would be accepted by them and subsequently by other authorities until it became scientific orthodoxy. No one person, no one organisation can decide what is a species and what is not. A species is more like an idea: and like most ideas, it can be modified, changed, argued over and rejected. So what is a species? Simple: the fundamental – or at least the most bleeding obvious – expression of evolution. Only members of the same species can mate, breed and produce reproductively viable young. I can hear whitethroat singing outside: the cock is trying to attract a female whitethroat so that the two of them can get together and make more whitethroats. You can mate a horse and a donkey but the resulting mule is almost always sterile. So not the same species. The trouble is that there are exceptions all over the place. Some animals go in for parthenogenesis, or virgin birth: reproduction without assistance from a member of the same species. Then there are examples of two different species that quite possibly could produce viable young but don't, because they never meet, being geographically separated. Conversely, two very similar species sometimes live side by side but keep separate, often sending out all kinds of strong keep-away signals to each other.† Similar-looking birds – whitethroat and lesser whitethroat, for example – will often sing markedly different songs.

A species, then, is a good concept but it's fuzzy round the edges. The idea of a species as a closed breeding community is OK: but suppose two populations of the same species start to behave in completely different ways and no longer have

---

* We did, however, come up with some important information for the Zambian Bird Atlas.
† In scientific terms the first is allopatric speciation, the second sympatric speciation. Readers may make their own Irish jokes here.

anything to do with each other? After all, that is exactly what's happening with orcas – killer whales – in the Pacific. Are they in the process of becoming two different species? Are we witnessing the process of speciation? Hang around a few million years and we'll have the answer.

So how many species are there in the Animal Kingdom? The number of species already described – species that are "known to science" – is usually reckoned to be a bit above a million. It's vague because ideas and methods of classification keep changing under the pressure of scientific examination, and also because, as I say, there is no one single definitive source and probably never can be. But it's certain that there are many more species awaiting discovery. Put a beetle expert* in the rainforest and he will find new species every day. How many species are there in total, then? I'm hanging on to my ten million because it's a non-contentious number with a bit of a bang to it. But even with the million-plus we already know about, we are lost. Already that is too many to hold comfortably in our minds: the mind-curdling numbers are just one more way in which we are alienated from the rest of life on earth. The numbers of our fellow primates are disturbing enough: wait till we get to insects, wait till we get to beetles, wait till we get to weevils. We think we are creatures of soaring minds that can encompass all of space and time, but the truth is that our own group of living things on our own planet is too much for us. So let's try to get our wheeling thoughts in some kind of order.

---

* A coleopterist: as such, closer to the concept of biodiversity than anyone else on earth.

Allspice, ant-killer

In Mary Tyler's rather good novel *The Accidental Tourist*, Rose alphabetises her kitchen. If you want ant-killer, it's next to the allspice. This is an unconventional method of classification, but perfectly effective. It is not without its dangers – you must be careful not to spice your ants or poison your hot-cross buns – but it's a fully functional taxonomy.

How do you organise your own stuff? I knew a man called Bob who loved motors. If a gardener has green fingers, he had black thumbs. His workshop looked as if anarchy had been loosed upon the world, but for him it was Order. He knew exactly where everything was: feeler gauge on the shelf up here, plug-spanner on the floor by the door, teacup on the workbench, well back. He could find what he wanted with his eyes closed, and no doubt often did. It worked brilliantly, but it was unique to him and therefore unshareable. If you wish others to use your system and to gain an advantage from it, you need an accessible logic.

Some people sort their books by colour, which is fine, so long as you know the colour of the book you are looking for and you don't have too many – or you don't need to look for a book very often. Some sort by size; most people have a special place for oversize books, like a coffee table. Some prefer a strictly personal Bob-like systemless system, in which recent still-to-be-read purchases are nearest to hand, old favourites always in the same place, and all the others are somewhere else. I used to sort my books chronologically: Homer near the start, Beowulf not long after, and

Joyce at the beginning of the end, but my wife found this irritating and made me change. Now, like Rose, I alphabetise. I keep the sports books separate; I have a lot because I also write on sporting subjects for *The Times*. I follow the same system with the sort of natural history books you read from end to end, but I keep books of species identification separate. That works for me, and makes sense for anyone else who looks for a book in my collection.

That's a basic taxonomy: that is to say, the organisation of things by means of their shared characteristics. And taxonomy in the natural world is no different. It's not right or wrong to alphabetise your kitchen; it's not wrong to sort out living (and formerly living) things in any way you choose. But what is classification for? Ultimately we classify things so we can acquire a better understanding of the things we are categorising. So that's systematics, and the way we systematise the natural world allows us to get a better understanding of evolution and of ecology. Ultimately, that is an understanding of continuity. Systematics reminds us that we – that is to say, every living thing – all share a history and a planet, and it tells us something of how this sharing has come about and how it works today.

The idea of conventional systematics is to group together living and extinct species that share a common ancestor: the common ancestor being the junction between branch and twig, between twig and twiglet: so we put humans next to chimpanzees as Rose put allspice next to ant-killer. We share a common ancestor with chimps that is not shared by mandrills; we share a common ancestor with mandrills that is not shared by bushbabies, and on and on and on, to the ancestor we share with nematode worms but not with oak trees.

Our taxonomy and systematics for the natural world are based on ancestors, not actions. Many species have evolved the same solution to life's problems, but got there by different routes. Bats, dragonflies, bluebottles and birds all fly: but they didn't all inherit flight from the same ancestor. They share a wing: they don't share a genealogy. You could group together all flying creatures in an alternative taxonomy, and preliterate societies often did precisely that. In an intuitive taxonomy, a fruitbat is much more like a bird than an elephant, and a dolphin is more like a fish[*] than a rhinoceros. Such a system works well enough: it's just not the one we use. We prefer a taxonomy based on evolution rather than function. Unlike Goldilocks, the taxonomy we work with is interested only in forebears.

This can be surprising, and frequently counter-intuitive.[†] If you look at the skies above Britain in summer, you will see four different species of flying birds with forked tails and swept-back wings: birds beautifully evolved for fast manoeuvrable flight. It is not exactly coincidence that they have the same sort of silhouette as fighter planes: though they hunt down and kill not enemy aeroplanes but flying insects. Of these four species, swallows, house martins and sand martins are closely related, sharing a very recent common ancestor. Swifts are in a quite separate group, for all their similarities of appearance and lifestyle. They got there another way, not via a common ancestor but by means of a convergence: same solution to the same problem by a different ancestral route. Intuitively, we view swifts and swallows as birds from the same shelf, but the taxonomy of inheritance places them a fair distance apart.

So far so straightforward, but taxonomy and systematics have been in a state of continuous revolution ever since it began with the great Linnaeus, who published *Systema Naturae* in 1735 and established for all time the mechanism by which we try and understand life on earth. Of late the pace has hotted up considerably. These days we use DNA

---

[*] I mean "fish".
[†] As this "fish" business makes perfectly clear – or perfectly obscure, anyway.

analysis to work out degrees of relatedness, when before we used observable physical characteristics. Some species we thought were very closely related turned out not to be closely related at all: it turns out that they are merely convergent. Other species thought to be far apart turn out to be close. This has been hard to adjust to even without the fish problem. There have been constant rows about human evolution and classification.

As you organise your library of living creatures, you have to decide what shelf to put each creature on, and for that matter, in which room of the sprawling library building. We no longer operate by the intuitively satisfying notion that life divides neatly in two halves – two kingdoms – of plants and animals. Some scientists recognise – and remember these things are constantly changing – 28 kingdoms of Bacteria and five kingdoms of the unicellular life called Archaea. There is a kingdom of Protoctista, which includes the amoeba, a creature most of us met at school. I was taught that amoebae were a frightfully simple kind of animal, famous for their ability to replicate by simple fission, splitting in half. Now they have been tossed into an entirely different kingdom from us humans – though I don't think that affects the celebration of the amoeba in "A Very Cellular Song" by Mike Heron of The Incredible String Band, great favourites of mine from my cosmic past and much loved still:

> When I need a friend I just give a wriggle
> Split right down the middle
> And when I look there's two of me
> Both as handsome as can be
> Oh – here we go
> Slithering and squelching on...

After the Protoctista we have separate kingdoms for fungi and for plants, which explains why gardeners, however green their fingers, are sometimes baffled by their inability to get an intuitive feeling for fungi and their needs. And after that – though there really is no after and no before, no order, no straight line and no ladder, for we are all, slithering and squelching amoebae and hard-talking hard-writing humans, viable and effective and fully evolved forms of life eminently suitable for the modern world – we get to the kingdom of Animalia: and here we shall stay.

Orang orang

Humans apart, I've never had much to do with apes. But an encounter with an orang-utan in Borneo has rather stayed with me. It was the eyes, you see. Or rather, the contact with the eyes. He looked at me and I looked at him. And it seemed to me that this eye contact was not without meaning. *Orang* means people; *orang-utan* means person of the forest in Bahasa Melayu, the Malay language. Malay charmingly doubles a word to indicate plural: public signs are always telling people what to do, and they do so by addressing us as *orang orang*. As a result, I can never hear the term orang-utan without being aware that it means a person.

How far away from our own species can we go and still have meaningful eye contact? You can gaze all you like into the huge and brilliant compound eyes of a dragonfly without ever getting a sense of the dragonfly responding to you as a person, as a fellow orang. David Quammen has written about his attempt to establish eye contact with a spider, something I would be unwilling to try myself, though I accept that this only exposes my limitations.

Eye contact with pet dogs is certainly meaningful: a frowning look will calm an exuberant dog, if he is basically on your side to start with. In the same way, a shared look of pleasure with a dog enhances a walk or a game. I am a horseman, and I know that eye contact with a horse is not a straightforward thing: two-way communication comes mostly in the form of body language and touch. I will instinctively avoid eye contact* in any situation other than one when I need to establish (or re-establish) dominance. With a horse, more than with most other domestic animals, eye contact is a threat, not an exchange of information.

I have frequently sought and given extravagant eye contact to baboons (who are monkeys, not apes) in camps in Africa, where they can be inclined to be overfamiliar. A hard Paddington Bear stare will cause them to back off and allow me to feel more comfortable. I have stared at crocodiles and felt the whip of danger, at a hippo and felt its irritation, at an elephant and felt a clear exchange of views on a temporary truce: if you don't come any closer, I won't either. I have never felt that same sort of thing with a bird, but all the same, there is an exchange of information: I have seen you and I know that you have seen me. This matters, but it is not recognition of yourself as a person. You feel spotted: you don't feel recognised.

That was the strange thing about the orang-utan. It was a big old male, and so he wore that curious and deeply unsettling face-plate or flange, which made him look as if he had just eaten a sandwich and had the plate for afters. It was unsettling because of a contradiction: there was something humanlike about the face, but the plate completely denied his humanity.

But I had a clear sense of recognition in that face – and what's more, I felt that the orang-utan felt the same thing about me. We are both, at least to an extent, face-readers: that's part of the way we both see the world, and it was very evident in this exchange of glances. Male orang-utans are pretty solitary: they lack the social skills and the facial expressiveness of the two chimpanzee species. All the same, an orang-utan knows what a face is and what it means.

---

* I suspect that is why horse people so often wear hats: traditional English cloth caps, cowboy hats, baseball caps: they allow you to drop your head and hide your eyes and so approach a horse without offering a challenge or threat. The more volatile the horse, the more important the hat.

The personal taxonomy of this big male informed him that there was something of himself in me, as it was clear to me that there was something of orang-utan in myself. And that's what non-threatening eye contact is: a recognition: an understanding that we stand on some kind of common ground.

Family feeling. That's what it comes down to. Nothing less.

My family and other family

The word "family" has a technical as well as an emotional meaning. This book is about the kingdom of animals. A kingdom is divided into phyla: we humans come from the phylum of chordates – backboned animals – along with birds, reptiles, amphibians and fish.* There are plenty of other phyla in the kingdom (your sexed-up garden slug belongs to the phylum of molluscs, as we have seen) and we'll visit them all in the course of this book – while remembering that there is no absolutely fixed and for all time agreement on how many phyla there are: some say about 20, others 36 and more. Classification is not definitive: it is about agreement, and science is not conducive to agreement. It's by disagreeing that scientists find things out. I must confess that when I was at school, I always thought that you could disagree about *Finnegans Wake* because that was subjective, but you couldn't disagree about science because that was objective. Then when I was at university they tried to make me understand about TS Eliot and the objective correlative as an approach to literature, while those studying science learned that their chosen discipline was a history of abandoned orthodoxies. Classification is like trying to tidy a house full of egocentric geniuses and small children: it's a great idea and you've got to keep trying, but you're never going to succeed in any final sense of the term.

The more scientists strain to reduce the complexities of

---

* Or whatever.

life into a single simple basic theory, the more complicated things get. It's life that buggers it up more than anything else; it is – relatively – and I use this word with some care – straightforward to come up with universal principles in physics. But life blurs all edges and every boundary. It was a British scientist, Ernest Rutherford, who famously said: "All science is either physics or stamp collecting." There is a pride in the bare, pared-down manliness of physics: the fuzzy, effeminate, indeterminate stuff called Life is forever shifting about on you. You can't rely on it. Even if you are as ferociously reductionist as Richard Dawkins, it is impossible to create – well, nobody has managed it yet – a fixed, firm and unshakable scheme for classifying the Animal Kingdom from top to bottom.

But I can at least tell you what it would be like if animals (and all other living things) weren't so confusing and so inimical to hard and fast definition. This ideal-world taxonomy is all about boxes within boxes, or to go back to your library, shelves within bookcases within rooms. Below phylum we have class: we belong to the class of mammals, slugs belong to the class of gastropods, along with snails, the creatures that inhabit most of the shells you find on the beach, and some other gaudy exotics we'll come to later. Next one down is order: we belong to the order of primates (the classification of slugs here goes slightly weird, so we'll skip that bit). Below that, we have family. Yes, indeed. There are 13 families of primates, and the one that includes us is the hominids. This group is divided into four genera (singular, genus): gorillas, orang-utans, humans and chimpanzees. The garden slug belongs to the family of Arionidae.

There has been a lot of alteration to the classification of hominids in recent years. Genetic studies show that we can't keep our fellow apes as far away from us as we would like. Originally, three species of great apes were listed as pongids, leaving humans refreshingly separate as hominids.

But this doesn't stack up genetically. It is now clear that there are two species of chimpanzee: chimpanzees and bonobos or pygmy chimpanzees. Both of them are more closely related to humans than they are to either gorillas or orang-utans.

Thus humans and chimps are now gathered together in the subfamily (taxonomy regularly and recklessly throws in sub- and superfamilies and other twists on the basic set of boxes or shelves) of Homininae, leaving the gorillas and orangs on their own separate branches of the family tree. Some would put our three Homininae species even closer. We have 99 per cent of our DNA in common, perhaps we should all be in the same genus: *Homo sapiens*, *Homo troglodytes* and *Homo paniscus*, or, if you prefer, we should rename ourselves *Pan sapiens*: the sapient ape. That would suggest that chimpanzees are human: or to put that another way, that we are all chimpanzees together. Meanwhile, the slug is from the genus *Arion*, specific name *hortensis*, making him/her *Arion hortensis*. So that's how a scientific name works: first the genus, with a capital letter, followed by a special name for the species: the specific name, in fact. And in italics, because that's the convention of science.

So now we can dive into the depths of this book: now the two great circles can begin turning. We'll take turns: each chapter that deals with us vertebrates will be followed by a chapter dealing with them inverts: or to put that another way, it will deal with all the other phyla of us animals. The next chapter takes us among the spineless ones; the chapter after that brings us back to the beasts with backbones, and on and on, until we get to the end... which is to say, of course, the beginning: there being no beginning and no end...

Below the drop-off

It's incredibly easy to be overwhelmed by the power and beauty and wonder of our fellow vertebrates. There are thousands of ways to do it, and I've experienced quite a lot of them: wildebeest in the Serengeti, dolphins breaching in front of the boat, eye contact with a bear, a colony of bee-eaters, a stooping falcon, a gathering of crocodiles, a horizon-filling chorus of frogs, leaping salmons, being within touching distance of 12-foot basking sharks, watching the *passeggiata* in the Piazza Navona.

These are great thump-in-the-gut experiences: things that stay with you forever. With all vertebrates, we feel a sense of identification – it's impossible not to cheer like a football supporter as the salmon goes for the top – that is perfectly complemented by a simultaneous sense of separateness. These are the equal and opposite aspects of the way we understand the endlessness of all these beautiful forms. It's when we stray further from ourselves that this sense of involvement, of gasp-making instinctive delight, is harder to find. There's a lot to be said about a nematode worm and a termite mound, but on the whole, they don't make you go phwoar.

The best way of transporting that sense of passionate identification into the world of invertebrates is by pulling on a mask and jumping into the sea. Not just any sea: you need coral to jump in over. Not that I've ever liked any actual sea very much. I throw up on boats, while all water, even salt water, has, across my life, resolutely refused to hold me up. As some people are left-handed, so some are negatively buoyant. I'm a sinker: so I've never felt that languorous sense of ease in water. I've never been weight-less: I have to work quite hard not to drown. (This is just one aspect of the endlessness of forms even within a single species. We vary in our degrees of buoyancy. Had humans become an aquatic species, a specimen like me wouldn't have survived to become an ancestor… though sinking has its advantages. I can get into deep water with a couple of flips of my fins,[*] while floaters must kick away like mad, so if I wanted to live by gathering food from the bottom of shallow seas, I might have survival advantages that a floater lacked. In this endless variety, and the consequent differential in survival, is to be found Darwin's great idea and beyond that, the meaning of life.) Even in modern life, I have found sinking an advantage. That's because my great skill of sinking has made it wonderfully easy for me to cele-brate – to go phwoar at – the incredible wonders of inver-tebrate life.

Coral. To place your face in a mask and see this undersea world without distortion is a breathtaking experience, particularly when it is impossible to take a breath. The sight has you literally gasping. It becomes still more vivid when you kick up your heels, flip your fins and shoot – if you are also a sinker – straight down among this fabulous stuff. These colonies of tiny creatures create vast sprawling edifices, an ecosystem that has been created and decorated and made glorious by the work of a billion billion inverts. We shall return to the coral reef a little later, to consider not just the coral itself, but the extraordinary and melodra-matic spectacle of biodiversity associated with it. But for the moment, let us contemplate the vast structures that the coral polyps create.

I have never tried to crack the scuba thing; I have always relished that unencumbered entrance into this alien element

---

[*] Prosthetic, not, alas, a part of my own endless variety.

that comes with mask, snorkel and fins. You need not trouble your head with breathing and pressure and air, and the only mechanism you have to worry about is that of your own body. So down you go, most wondrously of all, certainly most eerily, barrelling over a drop-off, where the coral wall marks the end of the shallow coral-filled seas, a vertiginous lurch towards the blackness of the benthic depths of the ocean. I followed as not-very-far as increasing chill, increasing fear and decreasing breath allowed, and then turned fin-flippingly lung-burstingly ear-poppingly up again, back to light and life and beauty and coral. That wall, that last beautiful bastion of civilisation against the terrors of the real sea, had been created entirely by invertebrates – oh brave old world that has such inverts in it. Chordate chauvinism, be gone, vertebrate jingoism, go hang: here was something to praise and glory in: here was something to try and understand: here are the invertebrates in all their beauty and variety, to seek out across the seas and across the living earth, in desert and rainforest and in back garden: in endless forms most beautiful and most wonderful: and most weird and most frightening and most alienating: and all of them, every species, every genus, every family and every phylum, at some dark level below the drop-off of human comfort, related to us, part of us as we are part of them.

Lemurs and archbishops

To call a man "a real man" is normally intended as a compliment; to tell a woman she is "all woman" is usually intended the same way. Americans say that someone is "a real human being" and mean nothing but niceness. To be humane is an unambiguously good thing. When we practise altruism on a large scale we say we are humanitarian. To call ourselves what we are is generally a compliment, then. At least at species level. To call someone an animal is one of the direst insults in our vocabulary. We are men or women, and that is good, we are human, and that is good, we are animals – and that is bad. It is, however, true.

To call someone an ape is another insult; it is also the plain truth. We don't use other levels of classification as insults. Get off me, you mammal! We don't use them as compliments, either: she's all mammal – no, that would never do. We don't call anyone a primate, either, unless he happens to be the Archbishop of Canterbury.

But primates we all are, and there are at least 300 species of us. There is no single characteristic that defines a primate; instead there is a suite of them: forward-facing eyes, eye sockets, grasping hands, nails, fingerprints and large brains. You and I fit in there very snugly alongside gorillas and mouse lemurs. The mouse lemur weighs in at 55 grams, or a couple of ounces. Traditionally, the title of largest primate is given to gorillas; a silverback – dominant male – can weigh 250 kilos or 39 stone. But this ignores the fact that humans are primates. The heaviest primates on record are humans, who have reached an astonishing 500 kilos or 79 stone.

Dwarf and mouse lemurs form a family; there are 22 species of them. Some are no more than 5 inches long: furry little things you might take for rodents at a casual look. But they're one of us all right: nocturnal forest-dwellers, some with complex social lives, and they have as much right to call themselves primates as you or I.

I have encountered the most beautiful of all primates – present company excepted – in the forests of Tanzania, the black and white colobus, a monkey spectacularly marked and moving with an easy grace. I have seen spider monkeys hanging by their tails in Belize: as they move through the trees with five functioning limbs they take on the truly disturbing appearance of immense arachnids. I have heard the night songs of howler monkeys that rumble through the forest, and I have heard the dawn chorus of gibbons, the wild whooping that rolls across the canopy and through the mist, making it almost physically impossible not to join in.

I have seen the baboons of Africa, the big males lounging on termite mounds with the air of a lazy man smoking a cigarette. I must confess that I have never been quite at my ease with them. One of the pleasures of being in the great parks of Africa is the absence of human beings: as near relatives, the baboons seem like intruders. They leave tracks that look dismayingly like the hands and feet of small humans, and their droppings are not dung or scats but an unapologetic turd, not unlike the ones that you and I call into being. Why is it that animal droppings are at worst unhygienic but human excrement is disgusting? When there were rows about the readiness of the Athletes' Village at the Delhi Commonwealth Games of 2010, the reports about "human excrement" being found made it sound as if it was far more horrible than any other kind. With baboons on the African savannahs, I always feel a little like Swift and the Yahoos: as if baboons were caricatured humans beings calculated to bring out self-disgust. This is wholly irrational: humans have a lot to be ashamed of, but we have no need to be ashamed of our ancestry or our close relations.

We are not alone. We are part of the continuum, all 300 of us. Primates all: the mandrill, the angwantibo, the hairy-eared mouse lemur, the red-tailed sportive lemur, the golden-headed lion tamarin, the white-nosed bearded saki, the muriqui, the crab-eating macaque, the hoolock gibbon, the gorilla, the chimpanzee, the Archbishop of Canterbury and me. And you.

Spineless

So as you see, I have adopted the common practice of splitting the Animal Kingdom into two, in order to make the two great circles of this book: vertebrates going one way, and invertebrates going the other. The only snag is that there is no coherent group called invertebrates. It's a bit like classifying all the books in the world as novels and non-novels. Which is exactly what a lot of people do: books are either fiction or non-fiction. That means, as I scan round my own books, that the complete poems of Gerard Manley Hopkins, the biography of Lucia Joyce, David Beckham's *My Side*, *I Ching*, *Wisden Almanac 1999*, *How to be a Bad Birdwatcher*, the *Shorter Oxford Dictionary*, *Birds of Belize* and *The Origin of Species* must all be considered roughly the same thing. In fact, all they have in common is that they are not made-up stories (apart from the Beckham, perhaps).

"Invertebrates" is not a coherent taxonomic group. It's a catch-all term for all the animals that are not part of our own group: if you're not one of us chordates, you're one of them inverts. Molluscs and arthropods and corals have no more in common with each other than they have with vertebrates like us. The division is just a convenient way for humans to reduce the massive and the incomprehensible and the alien and the multitudinous to a manageable form. The fact that this manageable state is an illusion doesn't trouble us: to see slugs and ants and worms as all roughly the same thing – and therefore different from lizards and weasels and ourselves – satisfies a deep human need. We find it easier to classify life, to *understand* life, by adopting a succession of binary views: my family and the rest of the world: English people and all other humans; humans and all other vertebrates; vertebrates and all other members of the Animal Kingdom. There is almost a moral division here: we feel less moral responsibility for non-humans than we do for humans: less moral responsibility for slugs than we do for elephants. We squash a mosquito that is annoying us; we don't squash a kitten, however annoying.

Most animal life is invertebrate. More than 90 per cent of all known species are invertebrates; and no doubt more than 99 per cent of the unknown species. The term invertebrates sweeps up, in its anthropocentric (or vertebrocentric) way, the most successful and numerous phylum on earth, the arthropods, which includes spiders and crabs almost as an afterthought. Most arthropods are insects: most living animals on the planet are insects. We talk about the Age of Amphibians and the Age of Reptiles and the Age of Dinosaurs and the Age of Mammals: in terms of numbers and diversity, it has been the Age of Insects for the past 400 million years. It makes more sense, at least in terms of numbers, at least in terms of relative success, to divide the Animal Kingdom into arthropods and non-arthropods.

Invertebrates are wildly various in every possibly way. In size alone they range from microscopic things to monsters 14 metres or damn near 50 feet long, weighing 495 kilos or damn near half a ton. This great diversification of living things began around 550 million years ago, with the Cambrian Explosion. Before that time, most of life on earth was single cells, or single cells organised into colonies. Evolution then went on fast forward, and over a period of a mere 50 million years or so, multicellular life broke out in a gloriously inexplicable fashion. Vertebrates were just one of the many possibilities that came up in this explosion, this sudden, this, in geological terms, almost instantaneous detonation of diversity. Darwin saw the suddenness of this change as a potentially serious problem for his theory of evolution by means of natural selection. Its causes are still the stuff of argument and controversy, though Darwin's position is safe.

The rest of the evolution of the Animal Kingdom has been a series of refinements and developments on this single brilliant idea of multicellular life. We vertebrates are just one of these developments: roundworms and lampshells and sponges and molluscs and arthropods are aspects of the same great notion.

Long-jump gold medal

I was embarrassed the first time I saw a bushbaby. How can you be serious about an animal with such a vomit-inducing name? I am a serious student of natural history; I couldn't deal with this debilitating attack of cuteness. We picked it out in the trees of the Luangwa Valley in Zambia: two red-hot coals reflecting back from a beam of light; their sudden disappearance as the owner of the eyes leapt into oblivion. I was quite glad I didn't see it better. I don't go to the bush for cuteness.

Some years later, I got a rather better view. Yes, back in the valley again. In full view this time: the face staring back. The eyes were much too big. They looked as if they had been designed with cold-hearted purpose – like ET – for raising the cuteness response. The name is going way too far. It actually comes from one of the calls, which is disturbingly like the mewling of a human baby. But when they move, they are nothing like babies.

I was on a night drive. We caught the twin coals in the beam of light once again; and then the leap. But this time, the animal leapt towards the vehicle. And he performed. Resting bushbabies make cute pictures, to the point of nausea, but the truth is that these creatures have their being in movement. They are leapers. And they leap like a man: springing from the legs to land on their feet, after which they grasp the branches with their hands. In the air, knees bent, arms outstretched, they look like long-jumpers: like little Bob Beamons setting record after record. They are more agile than we are and capable of proportionately much longer leaps, but they look like one of us, and for the best of all reasons. They are relations.

They too are primates. There are currently reckoned to be 20 or so species of bushbabies among the 300-odd primates, and probably a good few more still to be separately described. They are hard to find, hard to study and hard to identify: they live in the dark and they all look pretty much the same. This one, so far as I could make out, was the brown greater galago, or *Otolemur crassicaudatus*.

And he put on a virtuoso display of leaping. They can jump 20 feet, 6 m, which is not bad for an animal that's only a foot long, 30 cm, without his tail. This one was more modest in his ambitions, since there were trees all around, but the leaps were still prodigious for his size. It is impossible to watch a display like this without cheering at every leap; call that the Salmon Reflex. After that, this generous and obliging creature leapt straight over the vehicle – and it was a Toyota Land Cruiser – to give a display of equal brilliance on the far side before vanishing into the tree-crowded night, which is his world.

Wonderful things, bushbabies, with a complex social life that is still being investigated. Various mysteries are being discovered, including the fact that a mother can raise twins sired by different fathers. They all live on fruit and insects, this species being inclined to stress the insects. They catch them with their hands, like slip fielders, for they are authentic grasping primates. They can see in what we foolishly call darkness: with their vision and their clever hands and their quite phenomenal leaps, they make a very respectable living, and do so very largely unseen by the noisy primate that chops down forests.

Architects of human culture

Mostly, we think of inverts either with indifference or with fear and loathing. We tend to meet inverts in normal life only when something has gone wrong. Inverts tell us that our attempt to civilise the world is incomplete. Inverts tell us of the irrefragable failure of humankind: spiders in the bath, cockroaches in the kitchen, flies on the ceiling, daddy longlegs in the bedroom, the terrible itch of mosquito bites. Perhaps the most hated inverts of them all are the wasps at the picnic. Here they come, smart as paint in their striped livery, rascals on the spree, bingeing on jam, squash, Pimms and beer, homing in on picnics and barbecues and pub gardens across the land.

Is there anything good about wasps? Are they sent only to plague us, to spoil our attempts at a good time, to upset us on those few cherished fine days when the family is all together? Is their entire evolutionary purpose to frighten children, harass holidaymakers and sting al fresco lovers on the bum? Before we start to dive over the drop-off into the world of animals without backbones, let us pause just for a second to consider the fact that this book wouldn't be possible without wasps. Nor would *Wisden, I Ching, Finnegans Wake*, the poems of Gerard Manley Hopkins or *The Origin of Species*. Nor any of David Beckham's autobiographies. Human culture might have taken a very different form without wasps.

We only meet wasps on those end-of-the-season occasions, times when an intensely social and desperately hardworking insect gets a free pass to go out and have fun. We only meet wasps when they have finished with their duties of raising grubs into adults and securing the future of the colony. This period of all-consuming activity ends up with big numbers of workers with nothing left to do but have a bit of fun before the autumnal chill kills them. There are eight or nine species of social wasps in this country, including the much feared but comparatively gentle hornets. Hornets have a fearsome sting but are more reluctant to use it than other wasps. They are a decent, even rather alarming size, with brown rather than black stripes against the yellow: handsome little animals. There are also around 240 species of solitary wasps in Britain; none of them has a taste for picnicking.

Social wasps begin their lives when the warm weather comes. The queens, sole survivors from the madness of the previous summer, wake from hibernation full of last season's sperm. Their job is to make new colonies: or to speak genetically, to continue the old colony. Each one will make a nest, lay eggs, forage and feed the first generation of workers. Once that generation is up and buzzing, the colony can expand its ideas. The new workers take over the job of enlarging the nest and feeding the grubs, which means that the queen can concentrate on laying eggs. The grubs are fed on other insects, but you won't see a wasp carrying prey back to the nest. A worker catches an adult insect, stings it to death, rips off the legs and the wings, and then chews up the rest, to be fed as goo to the growing grubs back in the nest. The same process, minus the wing removal, takes place when they catch caterpillars and other larval forms. Eventually there are enough workers to raise a generation of males and queens. These wasps won't help around the nest: sex is all that concerns them.

Thus a series of generations, about one every fortnight, works to ensure the colony's continuation for the following season. Once that's done, it's the end of term. Nest's out for summer, nest's out for ever. And so these insects, which live a richly complex life, essential predators on all kinds of plaguing and pestilential insects of others species, go out on their annual foray to get themselves a bad name.

But we should bless wasps every time we see one. Without wasps, the spread of knowledge across the history of humankind would have been desperately hampered. Because wasps invented paper. A wasp's nest is as exquisite a thing as you will see anywhere in the natural world. It is a glorious piece of architecture created from wood pulp and spit, chewed up and manufactured into – paper. The Chinese cracked the technique a couple of thousand years ago and the rest is – in every sense – history.

I ask you, then, to raise your glass rather than your newspaper to the wasps you see as you take tea or drinks in the garden in the summer. These late-season hooligans live a highly evolved social life, are essential predators in a complex ecosystem, they create a thing of genuine beauty, and without them, what would Shakespeare have written on? "I'll be waspish, best beware my sting," says Kate in *The Taming of the Shrew*. We've got it all wrong about inverts: we do wrong to distance ourselves from them, we do wrong to hate them, we do wrong to look away from them. Fellow animals. What's so special about a backbone anyway?

The lion, the glitch and the glove compartment

I have the instincts of a beast. We all do. And I'm deeply thankful, because without them I'd be a few old bones on the banks of the Luangwa River. Yes: this is my Lion Story; don't expect me to be brief. Let's say we are sitting round the campfire at Mchenja, a camp on the banks of that river, a few hundred yards from where it all happened. You can hear the stridulating of crickets, the tinkling of the reed-frogs, the occasional bleep from Peters's epauletted fruitbat, the prooping calls of the African scops owl, the clink of a bottle on glass as I pour you a drink while you hold yourself in resignation for this set piece. I used it without any exaggeration whatsoever in my first novel, *Rogue Lion Safaris*. But then it doesn't need any exaggeration, though the drink helps.

A good few years ago, I spent a couple of months at Mchenja, when my old friend Bob Stjernstedt was "running" the place. One day, there being no clients in the camp, we drove off for a spot of birding. The previous evening I had claimed over dinner that I could pick out the stallion from a breeding group of zebra within ten seconds. I believed that my horsemanly reading of equine body language gave me all the information necessary. So every time we saw zebras, Bob stopped the vehicle and I picked out my stallion candidate. We then peered pruriently at the undercarriage through binoculars until we had a firm diagnosis. Male. Definite male. Yes, definite male. I was right

53

way above chance expectation, which was deeply pleasing.

We reached a point on the river a good way north, and we found, I think, a red-billed teal, an unusual bird for the valley, so that was all very satisfactory. We had a picnic, a bottle of Mosi beer each. I remarked that the front bumper of the Toyota Land Cruiser made an admirable shelf for a beer bottle; no danger whatsoever of it tipping over while I was scanning the shoreline for waders. We then tidied up the bottles and Bob started the Land Cruiser. Or rather he didn't. But no big deal. "Pass me the wallet of tools in the glove compartment, Simon." It kept happening: one of the leads would slip off the battery terminal and we would be temporarily becalmed. Bob had it right in no time: we started up and cruised back, stopping to stare intensely at small brown birds and at the undercarriage of zebras. We were within a few hundred yards of camp when the vehicle went lame on us. Puncture. Another routine emergency. I climbed into the back and passed Bob the Tanganyika jack, a high-lift jack beloved of old Africa hands. "And I'll need that wallet of tools again."

"You never gave them back to me."

"I bloody did."

"You bloody didn't."

A longish pause. "It's all your bloody fault."

"My bloody fault!"

"You told me the bumper made a good shelf."

"I meant for beer. Don't tell me: you left the bloody tools on it."

"Yes. Well, we can't have driven far with them. They'll be on the riverbank where we saw the teal."

"Anything vital in there?"

"Wheel wrench."

"Have you got a shifting spanner? Maybe you could bodge the nuts loose with that."

"Certainly."

"That's all right then."

"It's in the wallet with the other tools… Well, we'd better walk, I suppose."

"Back to the riverbank? It'll take three days."

"No, no, to camp. I've got lots of spare tools at camp."

So, unarmed as we were, we set off on foot through the bush. Precisely as you're always told not to. Stay with the vehicle: except that on this backtrack, we'd have been there for ever; it was more or less our own private road. So off we went. Just about the first thing we saw was a lioness. She was lying flat out like a cat on a hearthrug and never so much as lifted her head. I loved her. So we altered course, making a dogleg to avoid her, and aimed straight at the river; once there we turned right and followed it towards camp. I could just make out the shape of the huts as we crossed the Chamboo, a tributary of the Luangwa, dry at this time of year. It was as I was climbing the far bank that it happened.

A nuclear explosion. Rather drastically localised. A Combretum bush detonated before my eyes and became lion. Huge, black-maned, deeply shocked and utterly furious. I was, I suppose, about a cricket pitch away from him. Twenty yards max, though it seems a lot closer in my memories and occasionally in my dreams.

But here's the thing. I didn't cut and run. I wasn't even frightened, not then. Because it was here that my beastly instincts took over. I did absolutely nothing. I locked. Just staring at this angry lion, me looking at him looking at me.

So here's how I bring the story to a close round the camp-fire: "And I looked into his eyes and it was like looking into a fruit machine, fight or flight, fight or flight, fight or flight, and in the end it came down jackpot. Flight. The lion spun on his hips, revealing balls like footballs, and he ran twenty yards to an eminence a little further away from us and from there he lashed his tail and snarled his fury at us. There was one of those lifelong ten-second pauses. And then Bob said: 'Definite male.'"

Let's go back to that long frozen moment. My stillness was exactly the right response. If you, dear reader, ever walk into an irritated lion, then I hope you do exactly the same thing. I rather think you will. I suspect that this response is hard-wired in us humans. It doesn't make intuitive sense: you'd suspect that every instinct in our bodies would tell us to run, to climb a tree, to move, to get away. But this was no country to outrun a lion, which is capable of a charge of 35 mph. And had I run, I would have triggered his chase response, and I'd have been caught in a few yards. Caught and devoured. But he wasn't hunting: had he been doing so I wouldn't have had a prayer, no matter what my reaction had been. In fact, he was sleeping off a prolonged bout of sex: that was the only deduction to be made from the set-up we walked into. (And when lions go in for sex they don't mess about; the great ethologist George Schaller counted one lion through 157 copulations in 55 hours.) He was angry, not hungry. So standing still and staring him down was the perfect response, and I wish I'd thought of it myself. It gave him the utterly fraudulent message that I was not easy prey. I outbluffed a lion, but only because he wasn't terribly bothered in the first place.

There are two important matters arising from this camp-fire story. The first is that we humans first walked the savannahs of Africa a million years ago and part of us still knows it: enough for me to come up with a wholly appropriate response without reference to conscious thought or study of ethology. My body did it without reference to my mind. There is a continuity between humans of the 21st century and the first humans to walk upright. At an unconscious level, we humans are used to being prey, even if the idea shocks our conscious 21st-century minds.

It is an instructive thing, being prey. You realise all at once that so far as lions are concerned, we are no different from the impalas and pukus and buffalos. We like to think that there is something wrong when a lion or a tiger becomes "a man-eater": that the animal has been wounded and can't hunt its rightful prey, or that some bizarre incident has given the animal a depraved taste for human protein. But obviously for a lion, eating a human is much the same as eating a zebra. In 1898–99, two lions preyed heavily on the humans who were trying to build the Kenya–Uganda railways. The project was run by Lt-Col John Henry Patterson, who in 1907 published an account of this called *The Man-Eaters of Tsavo*, in which he claimed the lions killed 135 people before he shot them. I once read an account of this most unfortunate railway delay in a newspaper under the headline: "The Wrong Kind of Lions."

"To kill Man is always shameful. The Law says so," says Mowgli in *The Jungle Books*, but when you look a lion in the eyes – when you are made to look a lion in the eyes because your body says you must – you become aware of some different truths. That killing and eating humans has never been anything special or different or out of the way for a lion. That human uniqueness is not as clear an issue as we have been taught. To meet an alpha predator on terms of intimacy is to understand a truth that great libraries of philosophy avoid. That is why the order of Carnivora has a unique fascination for humankind.

Brother sponge

It's hard to feel a sense of equality with a bath sponge. Not much good looking for eye contact: eyes are one of the many things that sponges don't have. All the same, they are animals just as we are: kin to the tiger and the termite, and not to the oak tree, the pumpkin and the mushroom. They can even move, sometimes as much as 4 mm or 0.16 inches in a single day. If they don't look much like animals to us, that only reflects our narrowness of vision. They are animals in that they can't make their own food, unlike photosynthesising plants, which means that they must take in plant or animal food from outside. They are animals in that they give out carbon dioxide and take in oxygen, unlike plants, which take in carbon dioxide for photosynthesis (the basic mechanism of their lives) and give out oxygen as a waste product (which is one of the reasons why things like rainforests are quite useful for us animals, even if we don't live in them). There are also crucial differences at the cellular level – plant cells have walls; animal cells don't.

There are maybe as many as 10,000 species of sponges, and most of them are not like bath sponges at all: they build structures of calcium carbonate and silica spicules, and they wouldn't be much fun in the bath even for those with the most vigorous tastes in ablution. There are only a couple of genera of sponges with entirely fibrous skeletons, including the Mediterranean bath sponge that has proved so useful to humans across the centuries. The Romans used them; they have been used as padding for helmets, portable drinking vessels and water filters, for cleaning tools, applying paint, even for contraception. People used sponges attached to sticks for wiping their bums, hence the expression "the wrong end of the stick"; in such circumstances you'd prefer to seize the stick by the unsponged end. We like this species of sponge so much that we had brought it to the brink of extinction by the middle of the last century: that is the human way. These once-common animals are now rare and imperilled.

They are not easy creatures to empathise with. So much so that they were at one time allotted an entire subkingdom to themselves. These days they are considered a phylum, Porifera, part of the Animal Kingdom just like us chordates. Their great gift in life is to do with the circulation of water: it is by shifting water that their food – bacteria and tiny particles – comes to them, and by the same means their waste matter is washed away. Their various shapes – vase-like, tree-like, fan-like – are designed* to maximise both the natural movement of the water and their own ability to shift it about with tiny whips.

They are mostly marine animals, though there are a few found in fresh water. There are species that can deal with the rigours of the intertidal zones, others that can cope with depths of 8,000 m. They have a body of unliving jelly, sandwiched between two layers of cells. They can be no bigger than a few centimetres; they can be 6 foot tall, a couple of metres, or the same measure in diameter. They reproduce

---

* But not by a designer. I could, I suppose, adopt a more accurate, not to say pedantic locution here: "Their various shapes have evolved in such a manner that the circulation of water is maximised." The trouble is that human language evolved – developed if you prefer – before the facts of evolution were understood. So we don't have good words for it: we don't have robust language for this shatteringly robust concept. If you write about evolution, you have a choice: to use words that imply a designer or a purpose, but trust that they will be understood figuratively, or to clutter things up by sticking to strictly scientific terminology. So it's a question of which is less wrong. This is not a doctoral thesis, so I have taken the figurative option. That means that some stuff you don't take literally. Metaphors are the way we understand the world: we are not literal creatures. It takes a special effort to see things literally: outside of rigorous scientific literature it is not helpful or useful to try and do so.

sexually – simply shooting the stuff into the water rather than indulging in any gymnastics – and are hermaphrodites, like the slugs we considered a few pages back. There are sponges that look like loofahs but loofahs are not sponges. They are plants, specifically the fibrous structure inside the gourd-like fruit of *Luffa aegyptiaca*.

The profile of Winnie-the-Pooh

A bear has two faces. You find that out if you get close enough to feel vulnerable. I saw the first face as the bear advanced along the river. I was in British Columbia, in Canada, on a platform that gave a good view of the river but did little to conceal the observers. I was on Gill Island, in Knight Inlet. It was autumn: the salmon were running, so this was the time when the bears were briefly and annually visible. As the salmon come to spawn, so the bears become fishermen for a few short weeks of gourmandising. I had been waiting for a couple of hours, absorbing the quiet forest, hearing the occasional call of ravens. A pine marten, the first I had ever seen – the American species is different from the European – crossed the river by means of a fallen tree, as nice a little – well, fairly little – carnivore as you could hope to see. But it was the nice big carnivore I was looking for.

Waiting for wildlife. If you acquire a taste for the wild, you do a lot of waiting. Convinced you'll be unlucky, convinced that the day will be blank, trying to convince yourself that pine martens are enough to make anyone's day. And certainly they are on most occasions, but perhaps not when you're in a forest with bears in it. And if I hadn't known there were bears there, the walk to the platform would have told me: well, you know what bears do in the woods.

But the bear came. He arrived as if now was his moment

– any other moment would have been absurd – walking onstage with the immaculate timing of a great actor who knows that the more modest and unassuming his entry the greater the storm of applause that will greet him. This was a black bear. What must it be like, I wondered, to live in a country with three species of bear?* Huge and beautiful and imposing and quite tremendously black. He was so all-consumingly bear-like, so unutterably and completely ursine that I could scarcely believe it. Nothing has ever looked more like a bear than this bear. He approached, with his almost comic rolling gait; a bear moves very much like a pantomime horse. And there he was face on: round, kind-eyed, with semi-circular ears stuck on as an after-thought. He looked like a child's drawing of a bear; he also looked quite tremendously like a teddy bear. What could be lovelier, safer, more cuddlesome than the creature half-waddling towards me?

But then he turned his head sideways and showed me his profile. Showed me his other face. The face with a step in it, the big square muzzle sticking out rudely and incongru-ously. You thought I was your friend, you thought I was your brother, you thought I was the comforter from your childhood: I am nothing of the kind. I'm a bear. A real one. Deal with that, if you can. A modern teddy bear has no muzzle. Older teddies were less compromised, but with the teddies of today the face is flattened, humanised, turned into a furry person: like a human only cuddlier. Real bears look exactly the same, but only face on. A guide told me that he once had a client who wanted to be taken into the forest "so that a bear can lick honey off my nose". Possibly more likely to eat the nose off the face and take the rest of the head for pudding. The most fleeting glance of a bear in profile tells you all that and more, and you don't need a degree in comparative anatomy to understand what you

---

* Canada has black bear, brown or grizzly bear and polar bear.

see. With carnivores, our human responses frequently come from our wild past.

This is no lovely furry honorary human: a bear of little brain bothered by long words, a bear from darkest Peru brought up by his great aunt Lucy, a bear that calls out "yubba-dubba-doo". This is a carnivore. Look at the teeth inside that muzzle. Not just their bigness and their sharpness, though these are not irrelevant matters. Look at the way they overlap. The way they create a pair of scissors on either side. This is the carnassial shear: the scissor-bite that defines a carnivore. Next time you are patting a dog or stroking a cat, run a finger – carefully – along the side of the mouth and feel the way the teeth work. Run your tongue round your own molars and premolars. They are designed as grinders, not slicers. The fact that most humans eat meat doesn't make us carnivores, or to be more precise, doesn't make us carnivorans, or members of the order Carnivora. Watch a lion eat, or, perhaps easier, watch a dog with a bone. They use the side of their mouths, like a movie gangster with a cheap cigar.

There are around 260 species of carnivores. The smallest is the weasel, sometimes called the least weasel, found in Europe, Asia, North America and North Africa; it can be as small as 120 mm in body length, less than 5 inches without the tail, and weigh no more than 30 grams, not much more than an ounce. At the opposite end of the scale, polar bears, the largest land-based carnivores, can weigh up to 680 kilos, 1,500 pounds or two-thirds of a ton (twice the weight of a Siberian tiger), and be 3 m or nearly 10 feet long. Some sea-going carnivores are even bigger; we'll meet them in a chapter or two. Feared, beloved, hated, admired, persecuted, prized, mythologised: emblems of kings and football teams:* the great carnivores are the most talked-about and

---

* Five of the 32 teams in American National Football League are named for carnivores: Lions, Panthers, Bears, Jaguars and Bengals (which are tigers).

sought-after creatures on earth. Our childhoods are full of bears and wolves, and we are taken to zoos to admire lions and tigers. No group arouses such passions: and it is a fact in any discussion of wildlife issues, when carnivores come in at the door, common sense jumps out of the window forgetting its trousers.

Neon Meate Dream of a Octafish

I listened to Captain Beefheart a good deal in my youth, and it's good stuff. Mad and disturbing, but certainly good. Perhaps something of an acquired taste, though. Certainly, the Captain came into his own when we wanted people to leave the flat and allow us to go to bed, He never failed us. His greatest album, the ultimate flat-clearer, the one so many of my guests found too heavy to bear, is "Trout Mask Replica", and it includes tracks called "Hair Pie (Bake 1)", "China Pig", "Ant Man Bee" and "Wild Life". The title of this chapter is also the title of another song on the album. I think a track called "Glass Sponge" would have added completion here.

But it's not an oxymoronic mind-twister, or even a joke. Glass sponges are animals, and perhaps the most long-lived of all of us. They are also among the most beautiful. You find them at the bottom of the ocean, in places where the water is cold and the level of dissolved silica is high. In these circumstances they can, over great swathes of time, construct their fantastical selves: a body based around a structure of silica; a creature that is effectively made of glass.

They start as free-swimming larvae before settling down – some of them after as little as 12 hours – to construct themselves as cylinders or vases, and create a thin, gorgeous and delicate skeleton: a fantasy of architectural sculpture; six-rayed star-like silica spicules that fuse together into an intricate lattice, one that often remains standing there in its austerity and beauty long after the death of the animal.

They all have a large central chamber. The most renowned of all the glass sponges is Venus's flower-basket, a fanciful name that shows how these obscure creatures can fire the imagination. They are filter feeders, like the other sponges: some place them in the same phylum; others prefer to separate them.

They outlive us humans so comprehensively that it is impossible to get accurate data about the length of their lives. There have been mathematical estimates based on their rate of growth: these give the head-spinning answer of up to 23,000 years, which is the right answer from the data collected, but thought to be impossible in practical terms. However, as we stand, an estimate of 15,000 to 23,000 years has been tentatively accepted. If it's correct, there are glass sponges older than Niagara Falls, older than agriculture, older than writing.

There are also reef-forming glass sponges. These have been found 30 miles off the Pacific coast of Canada and Washington State; they were thought to be extinct but have been rediscovered. They can form colonies hundreds of feet long and up to 15 feet, 4.5 m, high: quite substantial things to go missing but the ocean is a big place and not overfull of humans.

Creatures made of glass, building undersea walls, some of them older than recorded human history... perhaps you thought I was exaggerating, merely showing off when I claimed that the Animal Kingdom is weirder than we are capable of imagining. If so, I trust you have changed your mind. If not, don't worry. I've only just started.

2. GLASS SPONGE

Il buono, il brutto e il cattivo

Che brutto! There were a lot of Italian clients passing through the bush camp I was living in, and for them it was essential to find the right adjective. The marvels we put before them each day could only be assimilated with the aid of the perfectly appropriate Italian word. A lion: *che bello*! An antelope fawn: *che carino*, how sweeeeet! And a hyena was always: *che brutto*! How ugly!

Nobody likes hyenas much. Evil-looking things, the more restrained Brits would tell us. Hyenas are cheats, scavenging the leftovers of the noble lions, stealing the kill from the beautiful leopards and giving everyone the willies when they were caught in the spotlight with their grinning Halloween-mask faces. The bare skin of the face gives them a particularly malevolent appearance; it doesn't help when you explain that this is an adaptation that helps them to keep clean, a useful thing for a beast that spends so much of its time with its head stuck in corpses. I have witnessed a hyena stealing a dead pregnant impala from a leopard and then, when it was faced with a threat itself, it salvaged something from the situation by ripping the foetus from the belly of the impala and running off into the bush to consume it. You can view this as a remarkably intelligent bit of improvisation or the footage from a nightmare, and be right on both counts.

Hyenas are carnivores. They look quite a lot like dogs, but are not all that closely related. There are four species, including the curious aardwolf which eats termites.* The spotted hyena is the one we are most familiar with, from wildlife documentaries if not from experience on safari. Their glorious whooping calls – they are animals who like to keep in contact with each other – are one of the great sounds of the African night: a spine-tingler for the first-timer, and providing a sense of homecoming for those of us who keep going back – giving us the chance to sit the newcomers around the fire and regale them with improbable tales of the bush: hyenas will fearlessly enter a tent and *bite the face off* a sleeping man...

A hyena always has an air of being up to no good, of being caught in the middle of some episode which even he must admit is rather shaming... but he's going to do it anyway, so off he goes into the bush, grinning hard. These animals never show themselves to great advantage, moving away at an awkward-looking canter, the sloping back and elevated front legs transforming them in an instant from a creature half familiar to one almost horrifyingly undoggy, the enormous penis whipping about unmentionably between those low-slung hindquarters. And if it's a really good-sized animal with a particularly impressive penis, it's almost certainly a female.

Hyenas have one of the most extraordinary social systems that we mammals have devised. They are dominated by females, but as a badge of their authority, the females have adopted vast false penises. This is in fact a radically developed clitoris, and it comes with fused labia that form a

---

* Evolution doesn't run along straight wide roads: it is a mysterious and often elusive business. The aardwolf is not the only carnivore that doesn't eat meat: the giant panda has gone still further and become a vegetarian carnivore. It is not doctrinaire about this, and will take animal matter in an opportunistic way, but its basic diet is plants. Genetic evidence demonstrates that pandas share a common ancestor with bears, as you would expect from the look of them, but they have taken a markedly individual fork in the road. Pandas don't hibernate, they subsist on bamboo, and they have acquired a spur on the wrist bone that acts in the manner of a thumb and is used for feeding: a rather brilliantly jury-rigged version of the primate's opposable thumb that you wear on your own hand.

false scrotum. They are impregnated via this organ, and they give birth though it, for it encloses the birth canal. This isn't always a straightforward process; ten per cent of first-time mothers die while giving birth.

Their social life is rigidly hierarchic and they have big and complex brains which make their complex society possible. They live by the clan, and everything they do is tied up with the female dominance hierarchy that underpins it. This contrasts with the social life of lions, which always seems to operate in a haphazard and rough-and-ready way. Lions seem to be making it up as they go along, and are perfectly capable of breaking all their own rules. Round a kill, the male does what he wants and everybody else scraps and snarls for the next bite, rather than waiting for a preordained turn... though the alpha male lion will often tolerate a cub – but no one else – eating alongside him. Lions are madly social but not really on top of their own social lives, in a manner that reminds me of modern humans.

But hyenas have it all worked out. The boss female is in charge: her close relations all gain status from her, and so get more food. It has been estimated that the top female is 2.5 times more likely to raise successful young than any other female in the hierarchy. The most dominant male in the clan ranks below the lowest-ranked female. A hyena den is, against all expectations, a lovely sight: always busy and doggy and playful, and the pups themselves, all black, look as if they would be perfectly at home on your own hearthrug. *Che carino!*

The human habit of making up morality tales based on the non-human creatures all around is probably as old as speech. It is natural to assume that all these things exist to teach us rudimentary lessons about the good, the bad and the ugly. I know a worldly and witty man who has given up shooting – apart from at hooded crows, because they are "such evil bloody things". In Ian Fleming's short story "The Hildebrand Rarity", we learn that James Bond "rarely killed fish except to eat, but there were exceptions – big moray eels and the members of the scorpion-fish family. Now he proposed to kill the sting-ray because it looked so extraordinarily evil". We tend to make such nursery judgments because we all learned animal morality tales in the nursery, and subsequently we have seldom been encouraged to think about animals in a grown-up fashion.

But there is beauty to be found in the hyena. They are possibly the all-mammal smelling champions, with a sense of smell that some say is over a thousand times more acute than our own: not so much an improvement as a complete new sense and with it, a totally alien world-view. Their skulls are remarkable: powerfully muscled jaws and bone-crushing premolars that give them one of the most effective bites in the world. Bone is nothing to them; hyena droppings turn white, because they take in so much calcium.

They exist by the rules and traditions of the clan. In some circumstances hyenas are hunters, capable of running down prey in open country because they can pace each other and take turns at the front as they make their endless, tireless runs across the bush; in thicker country they are more likely to forage and scavenge alone. But even alone, a hyena is part of a clan: that is the core of a hyena's being. You could make a fable about the good hyena with her acute sense of social obligation, contrasting her with the nasty anarchic lions who scavenge the kill from a pack of noble hyenas... but that would be just as misleading, even if more accurate from a naturalistic and a natural-history point of view. As so often, the truth and the beauty of the creatures we live with are obscured by human traditions. We are raised on untrue assumptions. We are not encouraged to look closely or think clearly about non-human animals – because, after all, we know everything already. Don't we?

Sod the rainforest

If you want to get the hang of biodiversity – to blow your mind with the impossible dizzying bewildering hallucinatory acid-stoned meaning of the teeming profusion and impossible variety and the endless numbers of species and species and species – don't go to the rainforest. Not that a good chunk of rainforest is ever anything less than wonderful – it's just not a very immediate experience. Laconic whistles of piercing clarity from the distant canopy; a brief glimpse of a butterfly the size of a bat; the feeling of being sought out and eaten by an endless variety of tiny things – a rainforest birder will frequently work wearing a bee-veil – and a certain elusive sensation that there is something going on somewhere but you can't ever quite put your finger on it.

Sod the rainforest. Let's go back to the coral reef. You don't even have to be a sinker. Put a mask over you face and immerse your head for, say, 1.7 seconds. And that's enough. Swear, blaspheme, pray, according to temperament. And that single swift glimpse will tell you the things that a rainforest never can: nowhere else in the wild world is it possible to understand, in one lightning intuitive Zen I-was-enlightened instant, about the endlessness of the beauty of the endless forms before you. And it all comes from a symbiosis so beautiful and so perfect that it might shake the faith of an atheist.*

---

* Let me quote David Attenborough: "My response is that when a creationist talks about God creating every individual species as a separate act, they always instance hummingbirds or orchids or sunflowers and beautiful things. But I tend to think instead of a parasitic worm [*Loa loa*, see page 104] that's boring through the eye

72

This time, let's not worry about the coral. Instead, we'll turn our attention to molluscs, worms, crustaceans, sea urchins and their kind; sponges, tunicates, jellyfish and their kind; seabirds, turtles, and of course, a wild and ludicrous assortment of "fish": so extravagantly decorated and so impossibly various that they seem not so much like creation by a designer but creation by a designer label. And all made possible by the coral.

It is a hard conceptual leap to make: it is difficult for us to think of a wall as a kind of animal. But that's a coral reef for you. Corals are animals in the phylum of Cnidaria, which includes jellyfish, sea anemones and the Portuguese man-of-war. The top layer of any reef is a colony of tiny animals, each of them not unlike sea anemones – except that they secrete an exoskeleton made from calcium carbonate, which supports and protects them. They use feathery tentacles to reach out and capture tiny crustaceans and other small specks of life. They cluster in groups, and build on the bodies of the colony's dead members, so that the wall grows.

Many species need light to make their way of life work. It follows, then, that you find most corals in the photic zones of the oceans, the levels where light can penetrate. This is to allow photosynthesis to take place. Not that an animal requires photosynthesis, not personally; rather, this is where the miraculous symbiosis comes in. Corals use the energy of single-celled algae that live within the tissues of their body, which in turn profit from the food taken in by the coral's polyps. For this reason many species of coral live in warm well-lit waters, not only surviving but creating an entire ecosystem around them. These tiny creatures are

---

of a boy sitting on the bank of a river in West Africa that's going to make him blind. And [I ask them]: are you telling me that the God you believe in, who you say is an all-merciful God, who cares for each and every one of us individually, are you saying that God created this worm that can live in no other way than in an innocent child's eyeball? Because that doesn't seem to me to coincide with a God who's full of mercy."

world-builders: they make it possible for a community to live where they do, to find shelter and to branch out in so many absurd and glorious ways. It has been estimated that as many as 4,000 different species of fish are associated with coral reefs.

Corals don't only make walls. They will also make eccentric shapes that seem whimsically, almost humorously put together: antlers, organ-pipes, giant convoluted brains, cauliflowers, table-tops. Hold your breath and fly down among them: a Dali landscape submerged beneath the ocean; a great feast of unlikeliness. It's as if you were travelling in the Heart of Gold spaceship in Douglas Adams's *A Hitchhiker's Guide to the Galaxy*, with its Infinite Improbable Drive. The whales in custard conjured up by this craft seem no more improbable than the pipes and pillars and hallucinogenic wonders of the coral world.

There's not very much of it. Corals cover 0.1 per cent of the ocean floor, but contain 25 per cent of all marine species. There are also deep-water corals, and there are some cold-water species as well, but it's warm and shallow waters where they thrive, and where they create their worlds of wonder, where they provide this crash course in the endlessness of both form and beauty. Darwin was a great fan of coral reefs. His first scientific monograph, published in 1842, 17 years before he detonated *The Origin*, was called *The Structure and Distribution of Coral Reefs*. It was about the formation of coral atolls, and correctly predicted that you would find a base of bedrock beneath each lagoon.

Coral reefs also demonstrate the extraordinary speed at which evolution can take place, when there is opportunity. The Great Barrier Reef, on the continental shelf of northern Australia, has only been there for 20,000 years and yet it contains ten species of coral found nowhere else in the world. Coral reefs can grow as much as 3 cm horizontally in a year, and up to 25 cm vertically, but they don't grow above a depth of 150 m.

Coral reefs provide the planet's most vivid lesson in diversity. They also preach a powerful sermon about the importance of small things. Lots and lots of small things are just as capable – perhaps even more capable – of altering the world as a few big things. Much of our world is shaped and moulded by creatures whose existence we are scarily unaware of. You'd think this would teach us to be careful about what we destroy.

The half-and-halfers

I realise now that my dream of achieving some sort of completion in this book is on the line. The fact is that I don't want to leave the carnivores. I could happily look at carnivores for the rest of the space I've allotted to mammals, or the rest of the space I've allotted to vertebrates, and maybe steal some space from the invertebrates as well. I have so many peak experiences associated with carnivores, so many stories, so many ways of trying to beguile an audience. You need never be bored when there are carnivores around. It's easier to fascinate with carnivores than with any other group of non-human animals on earth. Carnivores are sexy. Carnivores are also intimate; it's carnivores, more than any other order, that we choose to share our lives with. We bring canids and felids into our homes, we take toy ursids to bed with us as children. We fear the great predators, but at the same time, we love them, and above all, we envy them. When I read *The Jungle Book*, I was never Mowgli. I was Bagheera.[*]

And there are so many stories still to be told. I haven't told you about the otters at my place. I haven't told you about the stoats and weasels I see regularly from horseback. I haven't told you about badger-watching, and the glorious moment when that wonderfully improbable striped snout emerges. I haven't told you about my lightning-swift

---

[*] The black panther: "Everybody knew Bagheera, and nobody cared to cross his path; for he was as cunning as Tabaqui, as bold as the wild buffalo, and as reckless as the wounded elephant. But he had a voice as soft as wild honey dripping from a tree, and a skin softer than down."

glimpse of a tayra, a South American giant weasel. I haven't told you about my encounter with a jaguar in Belize, or the tiger I saw from elephant-back, not about the many times I have watched – it felt more like taking part in – the night hunt of leopards.

But no, I shall be disciplined, and move on to the animals that lack the charisma that even the meanest and gentlest of carnivores owns. Rodents and bats must not be shouldered out by the possessors of feet that can make no noise; eyes that can see in the dark; ears that can hear the winds in their lairs, and sharp white teeth.* All the same, I can't let carnivores go without first looking at seals and sea lions. They used to have an order of their own, but now the pinnipeds are considered part of the order of carnivores – and I'm damned if I'm going to leave them out. A fully-grown elephant seal is the heaviest carnivore on earth, then, and it can only function because it is usually in the sea. There are records of exceptional males measuring 20 feet, more than 6 m, in length and weighing 4 metric tonnes, 4.4 imperial. But all pinnipeds, no matter what their size, are important creatures. They are part of the interface between humans and the rest of the Animal Kingdom: one of the non-human animals with whom we have a special relationship. In the very early days of the conservation movement, the wider public first grew aware of the fragility of the wild world with images of baby seals being clubbed to death. This is a practice that still goes on, incidentally. It just stopped being news. The hooded bully with the club raised over a big-eyed baby seal became a symbol of human power over the natural world, and the vulnerability of non-human life.

A trip to "see the seals" is one of the great seaside staples of British life. Blakeney in north Norfolk in the high season reminds me of a night spent out in Deep Bay in the old

---

* The Hunting Verse, from the Law of the Jungle, learned by all young wolves in *The Jungle Book*.

days of Hong Kong, when the sea was crammed with tiny boats, themselves crammed with Chinese refugees, sometime using ping-pong bats as paddles: but here, they are all tourists going out to Blakeney Point to see the seals. I have been there many times with my own family, and it's always been a treat.

Seals have a meaning for humans of all ages because they seem such incomplete things. They are uncertain creatures, evolutionary half-and-halfers in a manner that perhaps reminds us a little of ourselves, for we always seem physically inadequate when compared to our fellow mammals: not as fast as a cheetah, nor as strong as a gorilla, incapable of flying and without any miraculous senses. Seals are creatures of the sea, but hearteningly, they are woefully inadequate on land.

Sea lions,* eared seals, are a good bit more agile, but they aren't exactly cat-sure on their flippers. The others, the earless seals, must drag themselves about like monstrous maggots. They are creatures of the sea who are still tied to the land because, unlike their fellow mammals, the dolphins and whales, they must give birth on land. They are required to spend a good deal of their lives on land and that makes them vulnerable. They need the undisturbed beaches, they need places where the men with clubs never go. I have heard it suggested that these never-fully-committed creatures of the open sea are "still evolving", which is a pretty thought, but one that misunderstands the nature of evolution. Evolution is not about seeking perfection: the only goal any creature seeks is survival and becoming an ancestor. So long as seals are doing this, there is no evolutionary pressure on them to change. In other

---

* Sea lions or eared seals have small visible external ears, and limbs they can use for active movement on land. They include the nine much-persecuted species of fur seals. The earless seals have to drag themselves about on land, though they can hump along like caterpillars for short distances. They have no visible ears. The walrus is in a third family of seals all on its own.

words, this rum set-up actually works, and evolution does not close a show that's making money.

Another wonderfully winning thing about seals is the moment of transformation. The waddling slugs hit the sea and in an instant they turn into sub-aquatic bullets, rocketing through the water with ease, grace and languor, or floating vertically as they doze, rocked in the cradle of the deep with only their noses above the surface. The transition from hopeless to glorious in a single splashless dive is something to be envied: something we have always yearned for ourselves, especially when we leave the rocks of adolescence to dive into the stormy oceans of adult life.

Leopard seals are the most spectacular seals, creatures of amazing ferocity. The males are more than 10 feet, 3 m, long, with a spotted throat that gives them their name: fierce predators of penguins and other lesser seals. There is a record of a leopard seal dragging a snorkelling woman to her death.

But the Weddell seal is the ultimate member of this group. It is the world's most southerly mammal, and it lives mostly underneath the pack ice of the Antarctic. It is found all around the continent, including the McMurdo Sound, which is 77 degrees south. It is named for Sir James Weddell, a British sealing captain who also gave his name to the Weddell Sea.

Weddell seals can be big, too: males and females both reach 3.5 m. All carnivores live by their teeth: it's by their teeth that you define a carnivore. That's true of Weddell seals, but not just because teeth are handy things to have when catching fish, squid, crustaceans and the odd penguin. They are essential for the seals' lifestyle: they use them for biting ice. The breathing holes in the pack ice are the life-support system of the Weddells, and they keep these open by chewing them. It is an expensive business: their teeth wear out faster than those of the seals with other ways of making a living. Weddells don't live much longer than 20 years; other comparably sized seals live twice as long. But a toothless carnivore is doomed.*

Weddells can forage under the ice for extended periods; they can dive for 80 minutes and can get as deep as 700 m, or 2,300 feet. But they are air-breathers like ourselves and must come to their breathing-holes or drown. Antarctic weather is always trying to close them up: Weddells are always working against the climate to keep them open. They will rest in the water – warmer than land – in the classic seal fashion, vertical with their noses above the surface in the manner known as "bottling". They will also haul out for brief periods. But in the winter, it can reach -70 degrees celsius up top, with winds of up to 120 mph. Come in; the water's lovely once you're in. At the coldest times of year, every other mammal in the Antarctic moves north: only the Weddells remain. They can hunt in the dark – below the ice, in the 24-hour night – by using their vibrissae, or sensitive whiskers. They are mammals like us, and land-users like us, but they are capable of living in a place where a temperature of -1.8 degrees celsius is the ultimate luxury.

---

* I have, in fact, witnessed a toothless lion holding onto his place as alpha male of a very feisty pride by sheer cool means. I watched him steal a kill from a younger, much stronger and properly toothed male using a kind of confidence trick, bluffing him out with a show of authority he had no chance of backing up physically. He then mantled over the kill and proceeded to *suck* the meat, slurping up what nourishment he could. The point here is that this old male didn't have to kill for himself – and because of his still-continuing psychological dominance he was able to exploit the teeth of others.

Walking plants

Sea anemones also seem like half-and-halfers, but not because they are a little bit like us humans. No: they confuse our sense of what is right and fitting by appearing to be half-plant and half-animal. We have even named the whole group after a plant. They are hard to think of as one of us: they live their lives in one place and do so by wafting their petals about. We humans and the snakelocks anemone have a good deal more in common than we have not – but that's not something that strikes an intuitive chord. We have something in common with leopards, yes, and perhaps with the octopus, but surely not with these fleshy, flabby pseudo-plants. We define ourselves by movement. We live in a mobile and dynamic way. Our ancestors took the revolutionary leap of walking upright, and having done so they led mobile nomadic lives across the African steppes. Movement is our notion of being alive. We feel the deepest compassion for people deprived of movement; the idea of paralysis terrifies us.[*]

When we seek recreation we mostly embark on some sort of movement: going for a nice walk, jogging, swimming, cycling, riding horses, skiing. For us humans, movement defines us: *eo ergo sum*.

Sea anemones can move a bit. They can creep about on their basal discs, and they can let go of their rock and swim/

---

[*] James Joyce writes of paralysis to express his despair at both Dublin and the world in *Dubliners*, his collection of short stories. Joyce as child narrator writes in the book's first paragraph: "Every night as I gazed up at the window I said softly to myself the word *paralysis*."

drift with bodily contortions and rippling of tentacles. If they find themselves in a bad place, they can leave it: they are not fully committed to the stationary life in the way that corals are. They have a get-out clause.

But sitting tight is what they are best at. It is a way of life that works: they have radiated out into more than 1,000 different species. Most are around 3 cm, not much more than an inch, but there are monsters a couple of metres or 6 foot 6 inches in diameter. They live by waiting to see what passes by: when it's small fish or shrimps they will harpoon them with stinging darts and fill them with toxins. They then draw them down into their single opening, which functions as both mouth and anus.

Plants that eat flesh are the familiar stuff of horror-stories, but even the real life plants that eat insects only do so as a back-up to the process of photosynthesis. But sea anemones, even though they look like plants, are animals through and through. That means that they cannot survive without taking in nutrients from the outside. In this case, they eat other animals. Some species of sea anemone have a symbiotic relationship with algae, like their fellow cnidarians, the corals, but these fierce flowers of the sea are animals just like you and me.

It was as animals that I first saw them when I went snorkelling in the rock pools on Rinsey beach in Cornwall: sometimes seeing them as soft little buds, at other times fully open, their tentacles the colour of dried blood. Tough little things: they could survive exposure to the sun as they waited for the tide to turn and bring back the sea. I had never seen them on family holidays to Weston-super-Mare and Southsea: this first trip to Cornwall was wilder and more thrilling than those tame resorts, and like all thrilling things, a little frightening. I had swapped ice cream vans and donkey rides for this: it was a plunge into a wilder world. The sea anemones made that quite clear.

It was through sea anemones that I had first-hand

experience of the fragility of nature, and of the power we humans have over our fellow animals. In 1967, the oil tanker Torrey Canyon took a fateful short cut, nipping in between the Scilly Isles and Land's End instead of going the long way round. As a result, the ship hit rock, broke its hull and the cargo of crude oil came flooding out. Rinsey was on the front line. But it wasn't the oil that killed everything. The detergent they used to get rid of the oil did even more damage. It took about 40 years for the rock pools to regain the vigour they had in my days as a boy snorkeller. It's only in recent years that you can look into those pools and once again see the gardens of anemones.

The Torrey Canyon disaster was an event that gave the nascent conservation movement a vividness that no amount of sermonising could match. The images of fouled and oiled seabirds raised both compassion and anger. Everyone agreed that this must never, ever happen again. And it didn't – until the next time, and the time after that. There's a moral there, and not a very subtle one: when it comes to a choice between shortcuts and common sense,* humans will always make the same decision.

It was the sea anemones that told me so. I rejoice to see them back on Rinsey beach. They'll be there forever now – or at least until the next shortcut.

---

* Common sense is not so common. Voltaire.

Wimbledon champion

We have a sneaking tendency to see the wild world as if it were the Wimbledon tennis tournament: to imagine that life is about the emergence of a single champion from many. In both the men's and the women's singles, Wimbledon starts with 128 players. Every year, half of them experience the despair of being a first-round loser. Eventually, after six rounds, we have a final, and after that, a champion: and that, of course, is what Wimbledon is all about. Things work differently on Wimbledon Common, and everywhere else in the wild world. The problem is that we humans love a narrative, and it is a love that can bring us confusion. Wimbledon tennis tournament is an annual narrative: from a confused and confusing beginning involving far too many individuals that moves inexorably to a simple and mean-ingful end. From chaos comes order. From many comes one, but the greatest of them all: that's a pretty satisfying thought, an archetypal tale that enthrals half the country and much of the world every year. It's a model we use unconsciously as we try to understand the world we live in; our politicians, for example, are selected after a struggle against many. They may lack the qualifications to govern a country, but they certainly know how to win an election. This model – the seeking of a champion – is staggeringly inappropriate to the natural world. We like to think of the lion as the king of the beasts, the greatest of all the carni-vores. We like the idea of the elephant as the champion land mammal. We like above all to think of ourselves: the champion of the apes just as the apes are the champions of the mammals just as the mammals are the champions of the vertebrates just as the vertebrates are the champions of the entire Animal Kingdom.

We've not only got that wrong: we've got it upside down. The best model of success in the Animal Kingdom starts off with one ancestor and ends up with 128 champions. Or more. Sometimes many many more. I have already talked about the more-than-1,000 species of sea anemones: when we reach the insects we will see what biodiversity can really do when it starts to get serious. Out of order comes chaos, or in any case a multiplicity that causes chaos in human minds. Mammals can at least give us the idea of biodiversity in a manageable way. Looking at mammals is like beginning to learn about the vastness of space by considering the solar system. That's why I have put mammals at the beginning – or at least one of the beginnings – of this book.

Which brings us to horses. The champion herbivore, at least in human eyes. The family of Equidae has seven species in it: three zebras, three asses and the horse. There are wild horses – Przewalski's horse in Mongolia is a subspecies of the horse we have domesticated, and there are feral populations of horses, most famously in North America, the mustangs, and most numerously in Australia, the brumbies; there are reckoned to be 400,000 of them. All these, plus all modern domestic horses, are the same species. All these horses – from a brumby to Erhaab, a Derby winner I have patted, to Fred and Mary,* the shire horses I once ploughed a field with, to Sophie, the Shetland pony in my field at home, to Przewalski's horse, to the horse that carries the drums at the Trooping of the Colour, to my old mare Dolly Dolores out there eating grass in the rain – are members of

---

* Excuse this showing off. Erhaab was trained by the great John Dunlop, with whose wonderfully generous cooperation I wrote a book about a year – 1987 – in a racing stable; I met Fred and Mary as a stunt for *The Times* and with their help I left a series of genuine if not entirely plumb-straight furrows across the field.

the species *Equus caballus*. The equids are one family with just a single genus, *Equus*.

If that makes the horse the champion of the equids, the bovids can offer 140 champions. In the bovids you can see the great inverted Wimbledon tennis tournament that produces so many champions from a single ancestor – and that is the way the great game of evolution is most effectively played. Evolution isn't a system for producing one perfect thing: it works best when it produces many things, all with a different method of surviving. The common genes carried in the equid family have only seven different sorts of survival machine, as Richard Dawkins explained it in his reductionist way. The bovids have 140: a dramatic improvement; still, wait till we come to beetles.

The bovids include all sheep and cattle, both the wild and the domesticated species, and all the antelopes. The vast and imposing gaur, a species of Asian cattle that stands 2.2 m, nearly 7 feet, at the shoulder, is a bovid, and so is Sharpe's grysbok, an anteloplet you could scoop up and carry under one arm: *che cariiiiiino*! Probably the smallest is the royal antelope of West Africa; adults can be no more than 10 inches, 25 cm, at the shoulder, and the males carry horns an inch, 2.5 cm, in length. Bovids make up some of the most numerous large mammal aggregations ever seen on our planet. The American bison has long since been shot to buggery, but the herds used to fill horizons. Springboks used to form wandering herds of many millions over hundreds of kilometres in what is now South Africa and Namibia.

One example of enormous numbers of enormous mammals still exists. It can be found in Africa, with the wildebeest migrations. The most famous of these operates in the Serengeti and the Masai Mara, that is to say, Kenya and Tanzania, for wildebeest pay no mind to international frontiers, though the Liuwa Plains migration in Zambia has the cachet of greater remoteness. I remember sitting for half

a day at a waterhole on the Serengeti watching the wildebeest coming and going. It never stopped. It was like staking
out Oxford Circus underground station: endless streams
going in both directions, so close that all idea of personal
space had to be abandoned, but they somehow kept the
peace by accepting the necessity of the situation. It was a
rush hour that never stopped: rush day, rush existence: these
are animals that have their being in crowds and company.
I spent the rest of that day on an eminence watching the
wildebeest making military manoeuvres across a vast plain.
The sound of these herds is both glorious and comic. The
onomatopoeic name gnu is almost perfect: they sound like
vast frogs, constantly exchanging brief clipped croaking
sounds: "*Newp!*" "*Newp!*"

The antelopes are the champions of the bovids, and
they make up a subfamily that contains about 90 species
in 30 genera. You give them a habitat, they'll adapt to it. I
have seen klipspringer on the rocky outcrops of southern
Africa, creatures that work on point, like ballet dancers.
They are small, though not grysbok-tiny, 22 inches or 58
cm at the shoulder. They stand on the pointy ends of their
hooves, and are said to be able to fit all four hooves on an
area you could cover with a two-pound coin. That's great
for mountains and kopjes, not so good for marshland.
But there's another antelope evolved to fill that niche. In
the marshes of Africa, you find the sitatunga with its long
exaggerated hooves that splay out, spreading the weight
and enabling it to walk on soft ground without sinking.
It's a fairly sizeable animal, and very graceful too, reaching
1.5 m at the shoulder, and built in long slim lines of almost
self-conscious elegance.

Antelopes can produce some monsters as well. I remember
a fine day spent in a hide high in a tree above a waterhole:
and a long, quiet and beautiful session was brought to a
conclusion with an odd clicking sound that carried across
the bush. I knew what this was all right: eland, as imposing

an antelope as you can find. The males can reach 6 foot or 1.83 m at the shoulder and weigh a metric tonne, 2,200 pounds. The male that came towards me was every bit of that, pausing to drink his fill, staying in sight for 15 minutes or so. The clicking sound is one of the mysteries of eland life, thought to be a snapping of tendons and considered a communication device, like a bird's contact call, something that allows them to keep in touch in times of poor visibility. They can move in pitch black without losing each other. Reindeer, used to travelling through white-outs, also make a clicking sound. Elands are not every fast, so they have an immense flight distance. They can't afford to let you come too near. If you get within 400 yards they start drifting off. But, high in my tree, I was invisible that day, and so the eland came closer than he would ever normally dare to. It was as if I was momentarily relieved of the burden of the human condition.

Infernal agony of gelatinous zooplankton

"'Brandy! Brandy!' he gasped, and fell groaning upon the sofa.

"Half a tumbler of the raw spirit brought about a wondrous change. He pushed himself up on one arm and swung his coat off his shoulders. 'For God's sake! Oil, opium, morphia!' he cried. 'Anything to ease this infernal agony!'… There, criss-crossed upon the man's naked shoulder, was the same strange reticulated pattern of red, inflamed lines which had been the death-mark of Fitzroy McPherson."

"The lion's mane!" were the last words of Fitzroy McPherson, who looked, when he died, "as though he had been terribly flogged by a thin wire scourge".

Sherlock Holmes[*] was baffled for a while, but he eventually came up with the murderer, which he then proceeded to kill without remorse. The killer was not the temperamental and jealous Ian Murdoch after all, as many had suspected, but a species of jellyfish: the lion's mane. The lion's mane jellyfish is real, the largest of all jellyfish, possibly the longest animal on earth. Its umbrella can measure as much as 6 feet 6 inches, 2 m in diameter, and the tentacles that hang beneath it have been known to reach 120 feet, 37 m, which is longer than the blue whale.[†]

---

[*] The story "The Lion's Mane" is in *The Case-book of Sherlock Holmes* by Sir Arthur Conan Doyle.

[†] In 1824, a specimen of bootlace worm was found on the Scottish shore. It was 180 feet, 55 m, long: pretty impressive, but it probably got to that length by being stretched, rather than by growing.

It's not actually a killer, though plenty venomous. Conan Doyle, a doctor of medicine, was careful in the story to establish that the first victim, Fitzroy McPherson, suffered from a weak heart. The lion's mane jellyfish's preferred place of living is the Arctic, and it seldom comes below 42 degrees north; Conan Doyle brought in a freak storm to carry his jellyfish to Sussex. It can cause pain in humans: there was an incident in 2010 when 150 people were stung in a single day at Rye in New Hampshire: all thought to have come into contact with the floating bits of the same dead one.

All jellyfish have a majestically sinister reputation: alien creatures from alien places that lure humans to their doom and sting them to death. The Sherlock Holmes story taps into an ancient tradition of monsters from the deep: fearsome things like the sea serpent, the kraken and Grendel's mother (slain so courageously by Beowulf). By the 19th century these fantastic monsters were often displaced by real species, monstered up and made many times more deadly for the sake of a great tale: Moby Dick, the sperm whale, the lion's mane jellyfish here, the giant squid, and in more recent times, the great white shark in *Jaws*; we'll be meeting sharks in more detail later on.

You'd think no one would be able to last for five minutes in the oceans of our world, teeming as they are with all those beasts longing to kill us. These creatures have a primeval fascination for us, one that has grown, rather than diminished. The more the land was conquered and the more the ferocious creatures of the land were pushed towards extinction, the more they began to arouse our compassion rather than our fear and loathing. But the oceans have a mystery that nowhere on land can ever rival, so oceans became year by year better places for rousing our thrilling atavistic fear of the wild world. Jellyfish are fascinating to most of us not because of their complex lifestyle, but because they sting: and because they really can, on occasions, kill.

Jellyfish is an imprecise term. It is a generalisation rather than a zoological category: loosely speaking, it sweeps up the free-swimming stage of many of the cnidarians. There are, for example, up to 1,500 species of Hydrozoa, a class of cnidarians, but by no means all of them make jellyfish. The medusa stage of the Scyphozoa life cycle is what we mostly think of as jellyfish: the archetypal genus being the moon jellyfish, often seen near the coast, glowing like little submerged full moons, and moving by pulsing themselves through the water. Jellyfish are just one of the many kinds of sea creatures we call fish even though they aren't fish at all, even in the loosest sense of the term. The Americans sometimes call them jellies, or sea jellies, in an attempt to avoid this confusion; they have also been referred to, with more accuracy than elegance, as gelatinous zooplankton.

The fact that we love to exaggerate the danger and the toxicity of life beneath the sea doesn't mean it can't kill you. There are species of box jellyfish – from the Cubozoa class of cnidarians – that are genuinely lethal. They move better than most jellyfish, able to control both direction and speed to some extent. They have eyes, though not as we humans understand the term, set in the clusters on the side of the transparent bell, and they can move around obstacles and spot prey.

Three species in particular are seriously venomous: the sea wasp, *Chironex fleckeri*, and *Carukia barnesi* (not, alas, named, still less described, by me) and *Malo kingi*. This last is a mere cubic centimetre in size, with tentacles trailing another 3 feet, a metre, also known as the Irukandji jellyfish. This is said to be 100 times more potent than a cobra, 1,000 times more potent than a tarantula. Having said this, most people who are stung make a reasonably swift recovery, but fatalities certainly occur. So that's all right, then. The seas still possess an eternal threat: and that remains an important thing for us all.

Walking with lechwe

Old buffalos turn evil. They'll charge at the least provocation. A lone buffalo is the most dangerous animal in the bush. Best thing to do if you meet one is to climb the nearest tree – but make sure it's a big one. If you leave a leg hanging down, they will lick that leg *to the bone*... Always one of my favourite campfire stories, this one. None of these tale-tellers has actually met an intrepid old man of the bush walking around with a mere bone for a leg after an encounter with a buffalo, but as a story of the boundless menace of the wild world, it's a hard one to beat.

The African buffalo, sometime called the Cape buffalo, is a fine and impressive creature, one that functions best in herds of 500 or so. When at ease they graze close and cosy, with a lowing grunting homeliness, a fall and flop of dung and a rich, pleasant farmyard smell. You can always tell a place where buffalos have been resting up: it looks and smells like a cattle byre. But they are warier than domestic cattle, because even creatures of this size are vulnerable: the lions of the Luangwa Valley specialise in buffalos, making them the most ferocious and intrepid carnivores on the planet.*

As the old male buffalos get stiffer and slower, they start to fall behind the herd as it moves from one grazing lawn

---

* There I go, boasting about the ferocity of a species or a population. It's a natural human response. I am, irrationally, very proud of the ferocity of my lions. Even if they're not actually mine. But perhaps the prize for pound-for-pound ferocity should go to the smallest carnivore, the weasel, which fearlessly takes on prey – rabbits – many times its own size.

to the next. These old boys form gentlemen's clubs. They still look majestic enough, but a little less mobile than they once were. If you see a couple of them, making a herd of two, they will tend to run away from you; if there is only one, he is more likely to run towards you. That's because if you are under threat by a predator, even when there are two of you, you have at the very worst a 50-50 chance of survival. A lone buffalo knows that in some circumstances, the percentages work in favour of a charge. It's not temper, it's maths. But we like buffalos to be malevolent creatures rather than merely seeking to survive by playing the percentages, just as we prefer our gelatinous zooplankton to be deadly poisonous.

Buffalos are just one species in the impressive adaptive radiation of bovids, and it is among the antelopes that this radiation is most compelling. The antelopes' basically similar body plan has been altered subtly by the forces of evolution, producing a range of remarkably different species with many quite distinct ways of living. They have developed different social structures, from the impalas that love to be in herds to bushbucks that prefer to live as half of a pair. I remember the time I took a lechwe for a walk in Lochinvar National Park in Zambia, when I had the chance to view the uniqueness of the species at close hand: very close indeed, because the lechwe would permit the occasional stroke or scratch around the head and shoulders.

This was an animal that had been rescued as a fawn and hand-reared, with the result that it responded more to humans than to fellow lechwe. There was something slightly odd, even spooky, about that, because Lochinvar is heaving with lechwe. No one, least of all a lechwe, could miss them: but this one walked around the fringes of the vast herds and stuck close to us humans. There are thousands of lechwe in Lochinvar, making the most of the damp and marshy expanses in the floodplain of the Kafue River;

this was one of the four subspecies, the Kafue lechwe.*
This individual had acquired the habit of walking out with
humans in the manner of a dog, stopping occasionally to
graze and browse, waiting patiently while humans watched
birds or examined other aspects of the wild world. Tim
Dodman inherited the animal from his predecessor when
running a project for WWF. As we walked, it was pleasant
to see the animal's athleticism. It was female, and in this
species only the males carry the good strong lyre-shaped
horns. She was tall enough, a metre, 3 feet 3 inches at the
shoulder, and finely built, probably less than 100 kilos, 220
pounds. As you watched her step rather primly across the
bush, you could see that her back legs were longer than
the front ones, giving her a distinctive gait. You see this
sometimes in horses that are not yet fully grown; "a bit on
the leg," horse people will say knowledgeably when such
a youngster comes up at auction. But with lechwe, this
legginess is an adaptation that helps them to deal with the
marshy terrain, and allows them to make spectacular leaps
across areas of deep water. It's the lechwe eyes I remember
most vividly: large and soulful, in the antelope way – though
that, of course, is because their excellent vision is a survival
aid, not a poetic prop for human writers – and covered with
a lattice of whiskers. I have never found this commented on
in a reference book, but it is presumably there so that the
animals can run fast though thick reeds and grasses without
getting stalks in their eyes. And on more or less the last

---

* A subspecies is a population that is significantly different from other populations
of the same species, but not different enough to count as a separate species. It's
just one more part of the fuzziness of taxonomy. Birds are always being promoted
from subspecies to full species, and sometimes back the other way. A recent
example in Britain is the Scottish crossbill. It was a subspecies of the crossbill
until 2006, when it was agreed that the songs of the two populations were significantly different, enough to make them two separate species, the Scottish and the
common crossbill. This was fascinating stuff for taxonomists; for twitchers, who
have a more sportive relationship with science, it was a thrilling development in
their game. Overnight, those who had seen both subspecies had a new species for
their lists: an armchair tick, in the jargon.

moment of my stay in Lochinvar, a herd of several thousand, stirred by some danger, performed a massed charge-past – it took minutes to go by – hammering through the reeds and splashing across the open water to vanish into the reeds again. They have evolved, and that triumphantly, for the wetlands of the Kafue.

Some years later, I was in Kenya writing about a humanitarian project that had been adapted by *The Times* as a Christmas charity* and was visiting a spectacular health centre in the middle of Masai territory, run by a remarkable philanthropist called Anne Lurie. It was a privilege to be there, and I didn't really need a reward, but I got one anyway. On one foray away from the health centre, to observe the outreach team visiting HIV-positive patients, gallantly riding across the bush on dirt-bikes, we came across some gerenuks,† antelopes I had always longed to see. These are amazing things, dry-country antelopes as the lechwe are specialised for wet. They don't need to drink; they get all their moisture from the vegetation they eat. They are slim and elegant even by antelope standards – they make the lechwe look positively chunky – but it's the neck that makes them extraordinary. The head is perched on a neck of improbable length and elegance: they are like antelopes trying to be giraffes. Which just about describes their way of life: they can reach higher than any competing antelope species, and so they can take leaves that are beyond the reach of rivals. They not only have these exaggerated necks, but they can also stand up, with a vertical back, and browse in comfort on leaves 2 m, 6 feet 6 inches above the ground. With their antelope eyes and their unusual build, they are an affecting sight.

---

* This was Riders for Health, which supplies motorcycles to health workers in developing countries, and, crucially, teaches them maintenance as well. Good people. Africa is full of expensive equipment that no one knows how to maintain: the 4x4 with grass growing through it is a frustratingly common sight.
† Hard "g".

So much so that the Hollywood actor and big-game hunter William Holden shot one. He then examined his work close up and exclaimed: "I've just shot Audrey Hepburn."* It wasn't a joke: it was a moment of life-changing horror. It was also a classic Damascene experience: Holden became a conservationist, and set up the William Holden Wildlife Foundation, which runs an education centre for conservation in Kenya.

---

* Hepburn and Holden worked together on *Sabrina*, 1954, and they were reunited for *Paris When It Sizzles* in 1964.

3. GERENUK

Life in the round

It's an odd thing, but the larvae of starfish and the rest of the echinoderms look like the embryos of vertebrates. And in fact, this phylum is comparatively close to our own. It's no surprise, then, that like ourselves the larvae are bilaterally symmetrical: we are all left-and-right creatures. We vertebrates grow up to become bilateral adults, but the adult echinoderms take a radically different direction. They live life in the round. They go in for radial symmetry. Mostly in a five-part way: where Shakespeare wrote his plays in iambic pentameters, echinoderms go in for pentamerism. The classic starfish body plan involves five limbs.

And that simple fact seems to me to enclose all the mystery of life. As a boy I was in thrall to the mystery of echinoderms: the compellingly alien round-and-round way of living was in itself a sermon on nature's eternal inventiveness. I longed above all things to possess the skeleton of a sea urchin: the hard, hollow, spineless dome that a sea urchin leaves behind when the rest of its body has decayed and been consumed, a skeleton that makes some of the most satisfactory fossils, wonderfully pleasing to hold in the hand. You could buy table lamps made from skeletons of the larger specimens of sea urchin. Strange thing: I can't remember if I ever, in fact, acquired a skeleton for myself. Perhaps I did long after the obsessive part of the desire had waned: a satisfying Proustian experience of disappointment, no doubt. Certainly the longing was by far the most vivid part of the experience: and that, I suppose, is what collecting is all about. Acquisition is always something of an anticlimax. (Faust and Don Giovanni both said some-thing along the same lines.)

Echinoderms include groups that look superficially dissimilar, though that's true of most phyla, our own included. As well as starfish, and the even more agile and mobile brittle stars, the phylum also contains mostly stationary groups like sea cucumbers, sea urchins, sand dollars, feather-stars and sea lilies. There are around 7,000 species altogether, and they are the largest marine phylum that has never diversified into fresh water, still less the land. They are diverse *ma non troppo*, a relatively conservative phylum that sticks to what works. And within these limits they are genuinely remarkable.

This five-way radial symmetry is the basic body plan but they have no great problems in deviating from it. There are starfish with six and seven arms, some with as many as 20. Some of the lilies – which like sea anemones look more like surreal plants than animals – can have 200 petal-like arms. Echinoderms use imaginative and various techniques for feeding; starfish are active hunters, cucumbers roll about on the sea floor and filter-feed on detritus, basket-stars feed on plankton. Most echinoderms favour the sea bottom, from the intertidal zones to the abyssal depths. The floor of the ocean is their domain and they have come up with a series of highly effective ways of exploiting it. Many of these sea-floor areas are low on life, but echinoderms are very talented at making the place work for them.

They confound our understanding of life's possibilities in two quite different ways. In all living things liquids are transported via the vascular system: sap, in the case of plants, blood and lymph in our own case. Echinoderms have a vascular system that operates on water. Uniquely, they operate on hydraulics, a system that is brilliant, economical and highly effective. Why didn't we think of that? But if we had done so, it's possible that we wouldn't have been able to break free from the watery life: we would have been forever barred from the land. From such accidents does the history of life unfold.

The second, rather more enviable development of the echinoderms is regeneration. If they lose a bit of themselves, they simply grow it again.* A starfish can lose one of its five arms and grow another. Some species will intentionally detach an arm, from which an entire new individual will develop: a particularly dramatic form of asexual reproduction, though echinoderms are perfectly capable of reproducing sexually. Sea cucumbers can discharge some of their internal organs as a defence mechanism; after they've done so they grow them back while surviving on nutrients stored in their bodies. Sea urchins regularly replace lost spines: this explains how some species can operate with such long and brittle appendages. The idea of replacing bits of yourself when they grow less effective has long been a human fantasy, though not necessarily a comforting one. Wearing out is our human birthright: echinoderms don't see the world in the same way at all. Lose a limb? Plenty more where that came from.

---

* Some vertebrates can operate limited forms of regeneration: see the section on axolotls on page 394.

Do I know you?

We tend to find the idea of being mammals a trifle dismaying. It's not something we like to think about. But we find it comparatively easy to deal with the idea that we are *related* to mammals. We can see some kind of kinship with a dog, a cat, a horse, a lion, an elephant. But when we get to animals further away from our notions of what a mammal should be, we find it rather more challenging. The fact that some of our fellow mammals lay eggs, for example, and that a mammal can carry poisonous spurs, are rather troubling; we will meet both very shortly. Marsupials that give birth to little blobs and nurture them in a pouch also stretch our minds somewhat: it is hard to think of ourselves as second cousin to a grub. But the creature furthest from human notions of what can possibly be a fellow mammal is surely the naked mole rat, and I introduce them here so we can get a real idea of what diversity means – diversity even in the class of animals that we belong to ourselves.

The naked mole rat, and its relation, the Damaraland mole rat, are the only species of eusocial mammals. That is to say, they live like termites, which we shall meet on page 401. They are found in Ethiopia, Kenya and Somaliland and they live in colonies. Each colony has a queen: the queen is the only female that reproduces, though she may have up to three males to fertilise her. She can give birth to as many as 28 pups at once. The colony can hold up to 300 individuals and as few as 20; the average is 75. The naked mole rat is up to 10 cm, 4 inches, long and weighs around 35 grams, 1.2 ounces. Apart from the queen and the fertilising males, the rest of the colony is sterile. The smaller ones gather food and maintain the nest. Naked mole rats live on underground tubers. They look after them, so that they regenerate and provide food for months, sometimes even years. The larger workers defend the nest. The queen can operate for as many as 18 years, normally giving birth to a single litter every year. She will not tolerate other females who behave like queens and will suppress them violently. When she eventually dies there is a power struggle among the top female workers, and one will eventually take over, generally after a violent battle for the mastery, or rather the mistressy. She will then stretch out her vertebrae and so prepare herself for the task of giving birth and nursing young.

They are rodents, as you'd guess from their almost horrifically prominent teeth: they look like a pair of teeth animated by a pink sausage, for they are pretty well hairless. They live entirely underground and dig with their ever-replenishing self-sharpening teeth. They have powerful lips that form a seal behind the teeth so they don't accidentally ingest earth. They move as fast backwards as they do forwards: useful for tunnel-life, when you can't always make a three-point turn. They have small lungs but are very efficient at taking up and using oxygen: important adaptations for the oxygen-poor environment of a tunnel. Uniquely among mammals they are thermoconformers, not thermoregulators, which means they can't dictate their internal temperature as we do, and so they move around the burrow to cope with changes in temperature, shifting to cool parts when it's warm, and huddling together when it's cold.

Another oddity: they lack the chemical – substance P – that transmits pain from the skin to the brain. They feel no pain when they encounter acid or pepper-spray. They are also remarkably long-lived. They can live 28 years, longer than any other rodent, and have a strong resistance to cancer.

They also have an ability to slow down their metabolism in times of shortage. Work is being done to map their genome with a long-term view to improving the lot of humanity.

So here are two species of mammal whose way of life is weirder than we can imagine. And they are part of our own class in the Animal Kingdom. They are our own kind: and yet they are as alien as anything else we will encounter in these pages.

Flatworm, flatworm, burning bright

Blake asked his burning tiger: did he who made the lamb make thee? Earlier in these pages, David Attenborough – no less a genius – asked of the *Loa loa* worm: did he who made the hummingbird make thee? And now I must ask of the schistosomiasis-causing genus of trematode flatworms: did he who made the black and yellow flatworm *Pseudoceros dimidiatus* make thee? The phylum – or the various phyla, depending on what sort of taxonomist you are – of flatworms brings together the beautiful and the damnable in a manner that is disturbing and dramatic even by the standards of the Animal Kingdom. There are endless forms most beautiful, there are also endless forms most terrible. Sometimes the two classes coincide in fearful symmetry. Sometimes the beauty seems almost impossibly benign; at other times the creature that stands before you is so uncompromisingly terrible that every human being must struggle to cope with the very fact of its existence.

There are 20,000 species of flatworms so far described. They have no blood system and no organs for breathing; they use their entire body surface for absorbing oxygen. This is not a trick restricted to inverts; some species of frog do it too. It makes sense, then, to maximise the acreage of body surface for absorption purposes, so that's why flatworms are flat. The smallest species have no gut either and take in food the same way they do oxygen, by absorption; the larger ones have a branching gut that reaches all tissues.

To a taxonomist they are specially fascinating creatures because of their bilateral symmetry. The invertebrate groups we have already looked at in these pages, the sponges, corals, anemones and jellyfish, all demonstrate radial symmetry. We tend to rate bilateral symmetry rather higher, not least because it is the body plan we work with ourselves, along with most of the Animal Kingdom. Flatworms are about as simple a living thing as can exist with this split-down-the-middle symmetry. They have a head end, to use the term approximately. Experiments have shown that they are capable of learning and remembering. Some species are capable of regenerating if cut in two.

And some of them are lovely things, rippling and feather-like as they glide their way through the water, sometimes in sumptuous colours, iridescent blue, deep crimson, purple, black and yellow, speckled with gold, so that you wonder about a world that can produce, apparently quite casually, a creature so obscure to us humans, and yet so gorgeous. This unexpected beauty of creatures that can be bigger than your hand is somehow deeply reassuring: benign creatures in a benign world.

The Turbellaria are a free-swimming group of 4,500 species, and they generally make a living by scavenging among the detritus, though some do so by direct predation. They possess ocelli, light-awareness organs too basic* to be dignified with the name of eye, with which they can detect the direction of light. Some species have one pair, others have three, though some species have clusters of them. They are hermaphrodites: in some species their courtship involves a bout of penis-fencing; the loser takes the female role.

Most of the free-living flatworms live in water, fresh or salt. A few can operate in damp soil and beneath logs. A couple of planarian species have been introduced to countries

---

* Demolishing one of the creationists' favourite arguments, who ask: "What good is half an eye? An eye must have been created complete or it is no use at all." Any flatworm will tell you different.

where the imported giant African land-snail has gone feral and become a pest. These flatworms preyed on the exotic and destructive snails in a very satisfactory way; alas, they stayed on and have started eating their way through native snails: oh, what a tangled web we weave when first we start to mess about with the natural world. This is a pretty terrible error: but it's not as terrible to us as some of the things we find when nature has not been buggered about with at all. And flatworms can be truly terrible.

Some species release eggs into fresh water by means of an infected human. The eggs hatch on contact with the water, and release free-swimming forms. These infect snails. They then take on another form inside the snail, and after that, they divide. From the snail, they release another free-swimming form, and this infects humans and some other mammals. They enter through the skin, and then progress to yet another stage. After that they migrate towards the lungs, where at last they reach an adult stage. From there, they migrate to a favoured body part: it might be the bladder, the rectum, the intestines, the liver or the spleen, or they might stay in the lungs. They feed on red blood cells and produce eggs: some species as many as 3,000 in a day. They don't often kill their human hosts – that, after all, would be counter-productive, like killing off your own planet – but they make sure their host feels like shit forever. This is schistosomiasis, sometimes called bilharzia, and the problems it causes are not lethal but chronic. You can carry on living, but your life is always ever so slightly awful, unless you get treated.* The parasites can, however, be a factor in bladder cancer. When children are infected, the parasites can affect growth and cognitive development.

Bilharzia is not the end of the flatworm's powers of destruction. Liver flukes are also flatworms: they get inside

---

* This can be done with a single annual dose of the drug praziquantel. Research into a vaccine continues.

the systems of mammals, including humans, and they too feed on blood. So there you have it: creatures with lives of a beautiful and bewildering complexity; some species go through as many as seven different forms during their lives. The way evolution has worked to bring this about boggles the mind, and if these creatures were even remotely love-able – or at least a little bit non-disgusting – they would surely be used by creationists as infallible proof of the hand of a designer. It is, after all, much easier to accept that some clever mind-beyond-all-minds worked out this septenary life form than to imagine how the forces of blind survival could have produced something so impossibly complex and so improbably effective.

But these brilliant little parasites have seized on humans as hosts. As a result, instead of being perfect for the crea-tionist cause, they are – from a human-chauvinist point of view – pure evil. We must be wary of such easy moralising, though, unless we are ready to base morality entirely on human convenience: i.e. if humans like it, it is morally good, if humans dislike it, it is evil. Accept that and you must accept that a lot of things on earth are pretty damn evil, from wasps at a picnic to the *Loa loa* worm. And by the same argument, rabbits are entitled to view humans as beings of pure evil: poisoning them, gassing them and shooting them.

If flatworms were capable of moral thought, they might look on an uninfected human being as Adam and Eve looked on the Garden of Eden: as a wonderful undeveloped proposition, as a glorious blessing from a benign creator. In the flatworm-centric universe, all that is good for flat-worms is morally right: and therefore human suffering is an unambiguously good thing. And if we humans want to see flatworms as evil, then we have to work out what kind of moral choice they have. They have evolved (been created if you wish to work from that argument) to feed not only on but in humans. Here they swim: God help them, they can do no other.

The elephant in the corridor

Hear and attend and listen, O Best Beloved, and I will tell you the tale of Baby Elly, and how she got stuck in the mud, and of Mummy Elly, who got stuck as well: both of them stuck fast as lightning in the dried-up dry-season devastation of Kapani Lagoon, and both likely to die: to weaken gradually from exhaustion and heat and dehydration before being picked off by the carnivores. It was Christina Carr from Kapani who sent me a set of pictures, taken by Abraham Banda, both old friends. The elephants got stuck right in front of Kapani Lodge,* and the mud was setting like concrete all around them.

The pictures make up a lovely comic strip. First they get ropes around Baby Elly and start to free her. But, frightened, she refuses to leave her mother and gets stuck again. They free her again – and there's Christina shouting and waving and trying to hoosh Baby Elly back to the herd, where they're all watching from a discreet distance. But Baby Elly gets stuck yet again. So they try a third time and this time they haul Baby away from her mother – and now one of her cousins calls for her and she turns and runs for the herd and the lagoon echoes with screams of encouragement from the elephants. Mummy Elly, close to exhaustion, is doused with water to try and stop her from overheating. The team then tie ropes around her and attach them to a tractor. She is not so exhausted she can't work out that what is happening is good, not bad, and she starts to cooperate. Eventually she

* In the Luangwa Valley in Zambia, where else?

struggles free and staggers back to the hollering herd and her baby. And then all the helpers head for Kapani Lodge and Chetumbe pours Mosi beers all around as he has done for me on many occasions past.

Of course, the correct thing to do in these circumstances would have been nothing. Let them both die. Part of the natural process. But you can't actually do that, can you? Certainly the Zambian Wildlife Authority and the South Luangwa Conservation Society (SLCS) agreed, and they did all they could to bring about this happy ending. The standard ethic states that you should only intervene when the problem has a human cause: SLCS routinely use tranquiliser darts to remove snares and treat the horrible wounds they inflict. The case of the ellies stuck in the mud elicited a purely emotional response, and if this was the Sixth Form Debating Society I could argue that it was morally wrong to interfere with the elephants' slow and unpleasant death, depriving the poor carnivores of a meat bonanza. But I certainly couldn't make the argument stand up if we were all standing in that dried-out lagoon with the suffering ellies before us. After all, blood is thicker than water.*

We humans have a powerful affinity for elephants. I wrote about the rescue story in *The Times*: the following week it went global, with newspaper after newspaper running the story as a classic heartstring-puller. It seems that across the world, we see elephants as special. I have on my desk a small figure of Ganesh, the lord of beginnings, the remover of obstacles, at times the astute placer of necessary obstacles, the patron of arts and sciences and letters, the *deva*† of intellect and reason: one of the most important gods in the Hindu pantheon. He is also respected in Buddhism and Jainism. Ganesh, of course, bears an elephant's head: this tiny likeness was given to me by the Wildlife Trust of

---

* The Dutch ethologist and ornithologist, Niko Tinbergen, used that phrase when explaining his passion for gulls.
† A Sanskrit word for god, which can be used of any benevolent super-being.

India (WTI) when I went there to support their work on elephant corridors. Elephant corridors are like the elephant in the room, only different. The idea is to acquire land between remnant chunks of forest, joining them up again, so that elephants can pass from one part of the forest to another along their traditional routes, widening their gene pool and decreasing their conflicts with local people. The chief exec of the WTI, Vivek Menon, is inclined to stress the genuine intelligence and genuine emotions of elephants, and to refer to them as "near-humans".

I'm not entirely at ease with this myself. It seems to me that the idea that the nearer to humans you are, the more rights you have is dangerous, difficult and subjective. (Do bats, being less like humans, have a lesser right to survive? And lizards? And butterflies?) But elephants certainly capture the human imagination: in their gigantic selves we see a part of ourselves: an answering gleam; some sense that the human condition is not a hermetically sealed thing that locks us off from all other creatures. It is said that elephants know what death is: they give physical support to the weak and dying and appear to mourn a death in the clan. One of the most heartrending scenes I have ever seen was the reaction of an elephant mother after her calf had been taken by a crocodile. No one who was there could ever imagine that grief is unique to humans.

There are two living species of elephants, African and Asian, and they are the two surviving members of the order of Proboscidea; some prefer three species and split the African elephant into two, the savannah elephant and the forest elephant. (Odd to think that England was once home to woolly mammoths, which went extinct here as recently as 10,000 BC.) It is impossible for such huge creatures to live in the modern world without clashes with people. Elephants kill people and damage crops and get in the way of trains and road transport. But elephants still survive, not least because people want them to. Elephants could be portrayed as malign creatures, every bit as evil as schis-tosomiasis-causing flatworms, but they are both liked and admired by humans, and not only by those who live at a safe distance. Indian trains occasionally strike elephants on one forest line[*]: the drivers involved are invariably trauma-tised, having to deal with the guilt of killing God. Elephants are hugely inconvenient things to have about: but – to a far greater degree than you would expect – people are prepared to deal with that inconvenience.

In Kerala I spoke to a community of people who used to live right in the middle of an elephant route. One woman told me what it was like to be inside a house while an elephant was destroying it. "I kept very still and quiet." The community accepted an offer from WTI to move: WTI purchased rich land suitable for spice growing and built a village there, far from elephants. The villagers took up this offer with delight: they saw it as a no-brainer. I was there for the official opening: I had the honour of cutting the ribbon for the official opening of one of the new houses. I then went back to the old and now abandoned village: the rice paddies had already dried up and the scrub was invading. The process of replanting the place with forest trees had yet to begin, but the elephants were already crossing and recrossing, as you could see from the soup-tureen foot-prints and the great fat loaves of dung. Vivek was inclined to be boisterous: "Look at this land! This is my land and it is ruined! And I am so *happy*!"

Elephants attract attention, elephants stir the emotions. The designation of these places as elephant corridors concentrates the mind most beautifully. But the fact is that if a corridor is wide enough for an elephant or rather, for a herd of elephants, it's wide enough for everything else as well. The elephants are just the figureheads. But what

---

[*] The line runs between New Jalpaiguri and Assam in eastern Bengal. In 2010, seven elephants were killed in a single accident.

sounds better: wildlife corridors? Or elephant corridors? We are instinctively on the side of the elephants. We are capable of thinking beyond our own species, not just as an intellectual effort, but with our hearts and guts: from the bottom of our being. We don't want to be considered apes or monkeys, we don't want to be called animals, but consistency is for wimps. We are proud to be kin to the elephants.

The holiness of tapeworms

Jain monks and nuns filter their water so that they don't accidentally swallow a fly when they drink. They go barefoot and they sweep the ground before them, so they won't inadvertently extinguish a small life as they walk. They do not wish to kill anything, harm anything or insult any living thing. Tolerance of other views is just one aspect of this. William Blake famously said: "Everything that lives is holy."* Which brings us back to flatworms.

How reluctant would you be to extinguish the life of a tapeworm? Are you prepared to concede the holiness of a living tapeworm? Tapeworms make up the Cestoda class of flatworms; in the adult form they spend most of their lives in the digestive tracts of vertebrates. About 1,000 species have been described so far, and they can affect every single species of vertebrates: us and our fellow mammals, all birds, reptiles, amphibians and yes, all "fish" as well. Tapeworms look like a length of tape: they are flat, being flatworms. You can ingest them in undercooked or raw pork, beef and fish, or by means of poor food hygiene. To be more specific, someone who absent-mindedly scratches his arse while making you a cheese sandwich could pass on a tapeworm

---

* The line comes from William Blake's poem *America: A Prophecy*. Blake was very keen on America as a place where humans could make a new start, free from all the terrible errors that we had made in the old world and in the old society. I wonder what he'd make of the place today. The sunflower that turns its head westward, towards America, following the sun each day, was Blake's symbol of this yearning for a new place, a new start. It is important to accept that no such place exists in the modern world. That's as true for human society as it is for the non-human creatures we share the planet with. We can only work with the places we've already got.

egg by means of faecal contamination.* I'm sorry, this isn't a very nice chapter, but at least it's a short one. If we're going to take on all living things, we have to look away from elephants and hummingbirds every now and then.

Beef tapeworms can be 12 m or 40 feet long; others can be even longer, 30 m or 100 feet. Perhaps their length is the truly sinister thing about tapeworms: a tiny fleck of a beast like a liver fluke seems in a way to be fair game: a thing half as long again as a cricket pitch is something quite different. (Sir David Attenborough, in a television interview, talked about his realisation, on return from some trip to the wild world, that he had "a little friend".)

Tapeworms work by attaching themselves to the intestine with what one can loosely call the head. It's called the scolex, and it has sucking grooves or hooks. Once anchored, the tapeworm can spread out and enjoy itself, absorbing food through its flattened surfaces. There are no symptoms associated with possession by tapeworm, certainly no vast and voracious appetite, as in popular banter. A tapeworm's host might feel some discomfort, and, perversely, a certain loss of appetite. It can survive inside you for as long as 20 years.

The length of a tapeworm is a slightly confusing issue. Any worm of respectable length is actually a chain of beings called proglottids, each one of which has male and female structures, and can reproduce independently. So it probably makes more sense to regard each worm as a colony. A whole playground of little friends, in fact.

Life delights in life. That is the less-often quoted part of that line from Blake at the start of the chapter. Where you find life you can generally find more life. Life cannot exist as a monoculture, for all that humans seem to be hell-bent on trying that experiment. Life depends on life. Humans depend on corn and rice and cows: tapeworms (some of

---

* Experts in food hygiene call the process "turd-to-tongue".

them anyway) depend on humans. It is not literally true that every possible way of making a living has some creature exploiting it, but there are far more creatures filling strange niches than the human imagination could ever come up with. Parasitism is just one of them. It works because life delights in life.

Plan A for aardvark

You only see aardvarks when you're drunk. You may be told this when you are travelling in Africa, and certainly, I can provide experimental verification. The only time I saw an aardvark was on the way back from a party in the Luangwa Valley. In point of fact, I wasn't terribly drunk; the party was rather fraught, dominated by a loud American hunter, and I found the occasion unsympathetic. All the same, when we found our aardvark on our merry way home, I instantly doubted such sobriety as I possessed. The most improbable-looking thing you could ever see: there it was, caught in the lights and looking like a bad drawing of itself. Sometimes, at moments of extreme excitement, you feel that it can't really be you out there in the wild seeing so wonderful a creature. I felt that when seeing my one and only tiger; I felt that when gloriously close to humpback whales. But when I saw my aardvark, I doubted not myself but the thing I was seeing.

This singularity is not a superficial thing. The singularity of aardvarks goes marrow-deep; no, deeper, twining down into the aardvark DNA. There's only one aardvark. The species is the only member of its genus, the only member of its family, the only member of its order. We like to push the claims for human uniqueness even though we belong to an order with 300-odd other species of primate. Aardvarks can give us a lesson in uniqueness. Just look at their teeth.

If you are a mammalogist, you spend most of your time getting excited about teeth. You don't spend your time looking at mammary glands, not when you're on duty, anyway, because just about all mammals have them and by their nature, they're not things that vary very much. Mammary glands unite us: teeth divide us. If you want to know what sort of mammal you are considering, look at the teeth. A good observer will be able to tell you what a mammal eats from the briefest examination of a few teeth or a chunk of jaw.

Teeth matter very greatly to almost all mammals. We have warm blood, we have a high metabolic rate, we feed our young on milk. In other words, we have a very high output, so obviously, we need a very high input – and teeth are all about input. Teeth are how we prepare our food for digestion, often how we take hold of it, sometimes how we catch it and kill it. For a mammal, teeth are in the front line of the struggle for existence.

And nothing has teeth like an aardvark. They don't need to chew their food, so they don't need incisors or canines. They have continuously growing open-rooted cheek teeth, described as "two upper and two lower premolars and three upper and three lower molars in each half-jaw".* The teeth are not covered by enamel but by a "thin bonelike tissue known as cementum". All of which means that aardvark teeth are unique and therefore aardvarks themselves are unique. In the ant and termite world-view, the aardvark is surely the embodiment of all evil: each one will – must – eat more than 50,000 insects a night. That's 50,000 individual lives required to get an aardvark through 24 hours of life: and we think lions and tigers are ferocious.

Aardvarks are tremendously good at what they do. When it comes to termites and ants these creatures are pros. They are superb diggers: four digits on the front feet and five on the back, each equipped with a spoon-shaped claw. Long, thin tongue, and sticky saliva to flypaper up the insects.

---

* Quotation from *The New Encyclopaedia of Mammals*, edited by David Macdonald.

They have a stomach that grinds up food, so they don't need to chew it; that explains something of their idiosyncratic dentition. They are also immensely strong. One of our post-party group – I shall leave his name out as it's something he's not so terribly proud of, and he's a very solid citizen these days – leapt from the vehicle and set off after the aardvark. As it dived into a hole, he grabbed it by the tail and tried to pull it out. He was a pretty strong guy, built like a Zambian scrum-half, and he knew how to use his strength, too. He planted both feet on either side of the hole and used his weight and leverage action to pull the aardvark free. Slowly, bit by bit, the aardvark, resisting the pull, dug deeper and deeper, until his brandy-and-coke-fuelled tormentor had to give up.

Aardvarks are so good at termites and ants that their numbers are diminishing – but not because they are killing and eating all the termites. The problem is that they are so good at what they do that they have no room to be good at anything else. They have driven up an evolutionary *cul de sac*: and that's all fine and dandy so long as everything around them stayes the same. Hyenas, jackals, vultures, storks, geese, pangolins, bat-eared foxes and aardwolves all eat termites, but not half as well as aardvarks. But all these others take other food as well, while aardvarks don't. The very strength of the aardvarks has made them vulnerable. The extreme specialist cannot adapt to changing circumstances. This is a conundrum sometimes called the tender trap. Intensive arable farming reduces the number of termites and therefore the number of aardvarks. The aardvarks, who appear so early in the dictionary, are creatures with no Plan B.

Unkillable bears

I must confess I had an Unworthy Thought in the course of putting this book together. Temptation shook me as a terrier shakes a rat, as it says in one of the great Modesty Blaise thrillers.* I thought it might be a smart wheeze to stick in a couple of hoax chapters.†

I could, for example, devote a chapter to something gloriously unlikely but borderline plausible, like flying mice. And I reckoned people would swallow just about anything when it came to inverts, so I could describe, say, a miniature bear with eight legs that you can boil or freeze or smother with radiation and still not kill it. But on second thoughts, I realised that nobody would believe that sort of thing.

If the principal aim of a living creature is survival,‡ we are wasting our time being human. If we want to lead the world in the art of survival, we should have been water bears. These tiny creatures are the all-Animalia masters of survival. For a start, you can find them everywhere in the world. They can cope with the tropics, and they can cope with the Arctic and the Antarctic. They are microscopic. Most species are less than a millimetre in length. They have four pairs of stubby legs, and they were named by their

---

* From *The Impossible Virgin* by Peter O'Donnell, one of the 11 novels and two collections of short stories devoted to Modesty Blaise.
† This is not a double-bluff, I promise you.
‡ This may look at a cursory glance like a reference to Richard Dawkins. It is nothing of the kind: Dawkins's contention is that the survival of the gene is what counts, that the host body – that's you, me, the aardvark and the water bear – are just survival machines for our immortal genes. This, in fact, is basic scientific orthodoxy, but Dawkins expresses it in particularly uncompromising terms.

4. TARDIGRADE

discoverer, Johann August Ephraim Goeze, who thought that their rolling, shambling gait was rather bear-like. So he said each one was *ein kleiner Wasserbär*, a little water bear, and they make up their own phylum with around 1,000 species. This is the phylum of tardigrades, or slow-steppers. They are sometimes called moss piglets.

That's because the easiest place to encounter them is on a chunk of moss. They prowl these miniature forests in their ursine amble, armed with jaws than can pierce cell walls. They use them to suck the life out of moss and algae, occasionally out of tiny invertebrates. And this frankly uncomplicated way of looking at the world has made them brilliantly ubiquitous. They have been found at 6,000 m, 20,000 feet, in the Himalayas; they have been found a depth of 4,000 m, 13,000 feet, in the ocean. They have been found under 5 m, 16 feet, of ice. In a single litre of sediment, freshwater or marine, you can find 25,000 water bears.

They can survive almost anything. They have failed to die at temperatures close to absolute zero. Boiling is nothing to them: they have survived temperatures of 151 degrees Celcius. They can stand 1,000 times more radiation than humans. They have been sent out into a low earth orbit and come back safe and sound. And when things get absolutely impossible, they have one more trick up their microscopic sleeve. They can just shut down and wait for things to get better. They can do this for ten years. When there is no moisture, they go into a phase called cryptobiosis: hidden life. It has been described as the instant coffee phase: they become flecks – almost completely nothing – and yet when a droplet of water comes along, they become living things once more. Some species have avoided that messy sex stuff, and have gone in for virgin birth, for parthenogenesis: an exclusively female population laying viable eggs. Just about everything they do is remarkable. Or perhaps it isn't, because what they do is survive.

This chapter full of fantastic achievements reminds me of the werewolf's boast in *The Chronicles of Narnia*:* "I can fast a hundred years and not die. I can lie a hundred nights on the ice and not freeze. I can drink a river of blood and not burst. Show me your enemies." This speech is about deep and frightening magic, the sort of thing that confounds the reader's expectation of what is possible and what is not. But down there, under the wrong end of the microscope, the water bears are doing the same sort of thing as a matter of routine.

I could, I suppose, put in a werewolf chapter in this book and try and make you believe it, though I don't suppose I'd succeed. But here we have an entire phylum of creatures, creatures found all over the world, at either end, in the middle, on the roof and in the cellarage, and they perform the werewolf's miracles as part of their daily lives. There really is no point in making up stuff about the wild world. The wild world is better at invention than we are: it not only has an infinitely deeper imagination, it has been doing it for millions and millions of years.

---

* From *Prince Caspian*, in the chapter called "Sorcery and Sudden Vengeance".

Flying flashers

I have no option, then, but to write about flying mice. Admittedly they're not mice and they can't fly, but apart from that the description is exact. But they look an awful lot like mice, and while they can't fly, they can most certainly glide. They operate on the same principle as the better-known flying squirrels, which make up a quite different and unrelated group. There are two species of flying mice, the long-eared flying mouse, and the confusingly named pygmy scaly-tailed flying squirrel. They are among seven species in the family of Anomaluridae, of which only six can get airborne. The two flying mice species form the genus *Idiurus*.

The flying mice glide by means of a membrane between their front and back legs which they open out – rather in the manner of a flasher – when they take to the air. They are from the vast and various order of rodents. They are a bit bigger than the house mouse that most of us have encountered, however reluctantly, about twice the size in fact, measuring up to 4 inches, 10 cm, without the tail, and weighing up to 1.2 ounces, 35 grams. They have a long tail that doesn't have much hair on it, but it has raised scales and patches of scaly skin to grip branches. This is interspersed with long hairs, which make the tail rather feather-like when used in aerial action and stretched out for balance. They are found in West and Central African forests, and they are nocturnal. Not a lot is known about them, because they are hard to find and almost impossible to observe.

5. IDIURUS

Gerald Durrell managed to catch some by smoking them out of their homes in hollow trees, though his attempt at keeping them in captivity was, alas, a failure. The reason I know about flying mice – they don't make it on the nature documentaries, and they are seldom found in pictorial reference books – is because of his description.* "I have seen some extraordinary sights at one time and another, but the flight of the flying mice I shall remember until my dying day. The great tree was bound round with shifting columns of grey smoke that turned to the most ethereal blue where the great bars of sunlight stabbed through it. Into this the *Idiurus* launched themselves. They left the trunk of the tree without any apparent effort at jumping; one minute they were clinging spread-eagled to the bark, the next they were in the air. Their tiny legs were stretched out, and the membranes along their sides were taut. They swooped and drifted through the tumbling clouds of smoke with all the assurance and skill of hawking swallows, twisting and banking with incredible skill and apparently little or no movement of the body. This was pure gliding, and what they achieved was astonishing. I saw one leave the trunk of the tree at a height of about thirty feet. He glided across the dell in a straight and steady swoop, and landed on a tree about a hundred and fifty feet away, losing little, if any, height in the process. Others left the trunk of the smoke-enveloped tree and glided round it in a series of diminishing spirals, to land on a portion of the trunk lower down. Some patrolled the tree in a series of S-shaped patterns, doubling back on their tracks with great smoothness and efficiency. Their wonderful ability in the air amazed me, for there was

---

* In *The Bafut Beagles* Durrell, at his best, is one of the finest nature writers ever to have picked up a pen, and by a distance the funniest. His *My Family and Other Animals* has rightly become a classic, but there are more than 20 more, not counting his fiction, and they're all full of good things. This chunk is quoted with the kind permission of Lee Durrell. The Durrell Wildlife Conservation Trust and the Durrell Wildlife Park on Jersey still carry on the work of this pioneer conservationist.

no breeze in the forest to set up the air currents I should have thought essential for such intricate manoeuvring."

It is clear that the ability to cross the gaps between trees is a huge advantage for those that possess it. That's clear from the fact that this ability has evolved independently four times over in mammals alone. We have the flying mice and the rest of the Anomaluridae. There are 44 species of flying squirrels. There are two species of colugos, also called flying lemurs, though they are not actually lemurs at all – common names are often confusing and approximate (we have already established that a jellyfish is not a fish). There are also marsupial gliders found in Australia and New Guinea, the flying phalangers. These four groups do not share a close common ancestor: each arrived at the same solution by a different route: another striking example of convergent evolution. Lizards, frogs and even snakes have also evolved, and again quite separately, the ability and equipment for gliding.

Cans and cans of worms

A fellow sports journalist told me a story about his days on local papers. He had been covering the local non-league football club, just as I used to do myself. Most non-league clubs are small places in which everybody knows everybody else and we all go into the bar for a drink afterwards. After one match, one of the players left his group of teammates to approach the journalist. He stuck a thick index finger under his nose. "If you ever call me again what you called me in the paper this week, I'll beat the crap out of you." The journalist was mystified, because he remembered that the player had had a pretty decent game and that he had made of point of saying so in his match report. When he got back home he checked out what he had written. The offending word was "ubiquitous".

You don't get much more ubiquitous than a nematode worm. Practically every living animal on earth is a nematode worm. It has been estimated that 80 per cent of the animals alive right now are nematodes. You can find a million individuals in a cubic metre of soil. Nothing teems like nematodes. They are sometimes called roundworms, and they have a phylum to themselves, Nematoda. So far 28,000 species of nematodes have been described, of which 16,000 are parasites. Most of them are tiny: around 2.5 mm, 0.1 inches, in length is typical, though there are some free-living giants of 5 cm, 2 inches, and some of the parasitic ones are longer still.

They are called roundworms for the excellent reason that unlike flatworms, they are round. The roundness gives them a muscle-lined body cavity, and so, unlike flatworms, they have a digestive system with a mouth and an anus. The cylindrical body is covered in a cuticle, which they moult every so often. They have what is described as a "relatively distinct head". Heads are by no means essential in life, even animal life, as we have seen with corals and jellyfish. There's no point in being headist.

Nematode worms lack the unkillable qualities of the water bears, but they are the world champions when it comes to ubiquity. You really can find them anywhere. Fresh water, salt water, ocean trenches, mountains, deserts, every continent. It is reckoned that 90 per cent of all life forms on the ocean floor are nematodes. They have been found far beneath the earth: even 3.6 km deep in a South African gold mine. They can survive heat, drought and frost, and many are capable of shutting down, enclosing themselves in a cyst. When things go well they can capitalise with extraordinary speed: a single individual can lay hundreds of thousands of eggs in a day.

They parasitise many species of plants and animals, including humans. The dreadful *Loa loa* worm, Sir David Attenborough's argument against a sweet-natured creator, is a nematode. Hookworms, pinworms and whipworms are among those that infect humans. It has been estimated that a quarter of all humans on earth are infected by nematodes: a condition associated with a warm climate, poverty, poor sanitation and overcrowding. The commonest form is ascariasis; a few people in Britain are diagnosed each year, though it is always something brought in from warmer places. The infestation may not have any symptoms at all, but in more severe cases you find fever, vomiting and diarrhoea. Infection is generally caused by (sorry to bring this up again) faecal contamination of food.

Pets and other domestic animals are often "wormed" to get rid of nematodes. But the forces of evolution can create some bizarre and ingenious methods of survival. There is a species of nematode that parasitises the fig wasp, which is the only pollinator of figs. There is another that infests a species of tropical ant. The condition causes the ant to develop a bright red gaster – which is what we scientists call the back end. Infected ants are more sluggish and carry the gaster in a bizarre way. This singles them out, and they get eaten by birds who mistake them for berries. The worms are excreted by the bird and when all goes well, the droppings are gathered in by the same species of ant, in order to feed the larvae – and that keeps the cycle spinning along very nicely, at least for the worm.

Some nematodes parasitise plants and affect the yield of commercial crops. The root-knot nematode is a particularly damaging species. Here you have a dilemma: they can be killed all right, but only at the expense of all the other species of nematodes. And many of them are beneficial. You can buy nematode worms from garden specialists and set them out in your garden to control slugs and other unwanted animals.

Self-sharpening chisels

Every idea is a great idea if it allows you to make a living on this tough ol' planet. That is true for animals, for plants, for fungi and for the five (or so) other kingdoms, all of differing kinds of microscopic life. Quite often, the forces of evolution come up with a design that works so well that it can be used for more than one species. Every now and then, you find a concept that works for quite a large number of species. The combination of large brain relative to body size, grasping hands with opposable thumbs and stereoscopic vision works well for us primates, and as we have seen, there are about 300 of us when it comes to species.

Which makes primate design a pretty good idea, but not half as good as self-sharpening teeth. That is the stroke of genius that allowed rodents to radiate out into 33 families containing more than 2,000 species. One authority states categorically that there are 2,277 species of rodents, but as we know, these things are in a perpetual flux. Still, flux or no flux, that's an awful lot of species to hang from one basic design coup: more than 40 per cent of all mammal species are rodents. Naturally, it is the teeth that define their difference from the rest of us.

When we think of rodents we tend to think of teeth. Timothy Spall played Wormtail in the Harry Potter films, a character who spends 12 years as a rat. When he resumes human form, his prominent teeth make him look as if the transformation were only 98 per cent complete. Laurence Olivier played Shylock in *The Merchant of Venice* with false teeth designed to give him an appropriately rat-like face.

All real rodents live and die by their teeth. Their unique arrangement is in the incisors: the ones at the front, two pairs, one top and one bottom. Often orange or yellow. These teeth grow throughout their lives. The front-facing surfaces are conventionally enamelled; the back surfaces are not: they are exposed dentine. The teeth close over each other, and this creates a self-sharpening mechanism which means that the enamel edge is honed every time the animal closes its mouth. As a result, the edges are always as sharp as chisels.

Rodents gnaw. If they stopped gnawing their teeth would grow out of control. Small rodents gnaw small stuff, on the whole, while big rodents gnaw big stuff. Beavers gnaw down entire trees and change their local environment as they do so. This gnawing ability gives rodents a great advantage: they can eat seeds in hard cases, food unavailable to less well-endowed animals. They can gnaw through all kinds of things to reach food. They also use their teeth in self-defence if they have been unable to pre-empt confrontation by running away. The image of the cornered rat is one based on fact; all predators prefer easy prey, because they are not in it for the sport. An animal that makes itself painful to kill is giving itself a considerable advantage over a mere fleer. Rodents also have a clever device to stop them choking to death on wood splinter or seed husks: they can suck their cheeks into the gap between their self-defining incisors and the comparatively mundane cheek teeth set further back in the jaw.

So much is common to all rodents. The disparity between species is equally remarkable. Rodents live underground, on the ground and in tree canopies never touching the ground. They can live in deserts and never take a drink, they can live in water, or semi-aquatically at any rate, and some species have webbed toes to help them do so. They can be found on every continent, including the Antarctic, because some species have grown adept at living off and alongside humans. Rodents have been found living in cold-stores – not just raiding, but living full-time inside them – and at the bottom of mines. They include – to take a lightning and rather selective run through the order of rodents – beavers, capybaras, flying squirrels, non-flying squirrels, dormice, voles, chipmunks, mice, rats, porcupines, chinchillas, guinea pigs and coypus.

They usually have a pretty small body size: the African pygmy mouse is 6 cm, 2.4 inches, long without the tail, though the Baluchistan pygmy jerboa can be even smaller. The biggest is the capybara: a majestic giant guinea pig. I have walked among a gathering of them in a public park in Curitiba, in Brazil; I felt like Alice after she had drunk from the bottle that said Drink Me and shrunk drastically, as I strolled thigh-high through guinea pigs. Capybara can reach 65 kilos, and an exceptional animal weighed 91 kilos. But these are midgets compared to some of the extinct species of rodent: *Josephoartigasia monesi* was, it has been estimated, capable of reaching 1,000 kilos, 2,200 pounds or more.

Rodents are mostly compact and short-legged, but where there is an absence of antelope competition in South America, you find long-legged, rangy rodents like agoutis, maras and pacas. Rodents don't have the complex ruminant stomach that antelopes possess, which allows them to digest cellulose and so get enough nutrition from grass. But rodents who specialise in eating greenery manage the same trick by passing it through their system twice: that is to say, they eat their own droppings. The second time through, the droppings are hard and unpalatable. Rabbits (not rodents) do the same, which is how they function so effectively as grazers: if you have read *Watership Down* and wondered what Richard Adams meant when Hazel and Fiver were relaxing in their burrow "chewing pellets", you need wonder no longer.

Rodents have other advantages that have contributed to

their extraordinary success. They are small, and they have a rapid breeding cycle. In optimal circumstances, they can reproduce at a rate that pest-control people love to scare us with: six mice in a house can become 60 in three months; one female can produce up to ten litters a year, each of a half-dozen young, all of whom could be breeding within 30 days.

Humans tend to hate mice because they invade our homes and eat our food and foul our cupboards and lofts and damage the places where we have our being. Perhaps it is helpful to think of humans as mice and the natural world as the home of others species: a home that we invade and foul and damage. We are probably more capable than mice of making a moral choice: pity we don't do so all that often.

Tipping the velvet worm

A woman walked into a bar and asked for a euphemism. So I gave her one.

Rather a favourite joke of mine, that, and it came to mind with the requirement to write a chapter on velvet worms. No doubt it got there at least partly because of a 1998 historical novel about a lesbian love affair. It was called *Tipping the Velvet*, apparently a Victorian euphemism for cunnilingus; the book was made into a television series.*

But both velvet and worms have long been used in sexual codes, so you'd be entitled to think that a velvet worm chapter was an elaborate practical joke, if not a rather diseased and unhealthy fantasy of the author. Nothing of the kind, I assure you. Velvet worm is the common name for the phylum of Onychophora, which means claw-bearers. It's not one of your big phyla, no more than a couple of hundred species described, but they are related to the biggest of them all, the arthropods, which we'll come to in due course. It has been speculated that velvet worms are relict ancestral arthropods, so you can look on them as both a failure and the greatest success in the Animal Kingdom. Water bears, incidentally, are also related to arthropods. Rum the way it all works: the same branch of evolution produced these two obscure phyla which between them have only a quarter of the species of us mammals, and we're not even a full phylum. But the third

---

* Such matters are (a) more acceptable and (b) more titillating in period costume.

turning on the same branch provided the arthropods: and practically every living species of animal is an arthropod.

It's not exactly a question of where did it all go wrong, as the bell-hop so famously asked George Best.* It's more a demonstration of the truth that one basic design is more suitable than others when it comes to producing endless forms. It's a nonsense to speak of dead ends, or of failed experiments. Velvet worms, like water bears, are viable and effective animals. It's just that they're not very various, or for that matter, numerous. They work, though: and that's enough to be going on with. Life is not about being ambitious; it's about continuation, and velvet worms continue.

So far as humans are concerned, velvet worms have their being in obscurity: making a life in leaf litter and crevices in warm parts of the world, especially rainforest, in the Americas, Australia and Africa. They are mostly nocturnal, though they are often active after rainfall. They are hard to spot: you are unlikely to find yourself out in the forest with a guide keen to show you his velvet worm.

They look like caterpillars; some more like giant caterpillars. They have bodies a little like earthworms, but with stumpy legs that make them look a little like millipedes, though they are related to neither. They are, to use a delightful phrase, "obscurely segmented"; in other words, their body is divided into segments (like insects and other arthropods, in fact) but the segments are quite difficult to make out. This is because they are not rigid animals; they are kept firm by their fluid content. They can dry out in adverse conditions, like other non-related creatures such as worms and sea anemones. They need high humidity in order to be safe, which makes rainforests an immensely suitable place for them. They stretch and contract their

---

* Apologies if this ancient footballing story is unfamiliar. The incident took place after Best had retired from football. On the day in question he had taken to a hotel bedroom after a sizeable win at a casino. The room was full of fivers and a lovely lady, usually described as Miss World.

bodies in order to move, and they do so on stubbly little feet with the pleasant name of lobopods. They can be anything from half a centimetre to 20 cm, a quarter-inch to eight inches. Most of them are reddy-browny in colour, but there are some decorative ones in bright green, blue, gold and white, and some are patterned. They move in a slow and graceful fashion as if they were doing an invertebrate form of tai-chi, which makes it hard for predators to find them; most predators' eyes are tuned to catch movement. All the same, velvet worms are highly effective ambush predators, and they use a single terrible weapon: slime.

Their basic strategy is to lie in wait for a small creature and then to slime them. They can squirt this as much as 4 cm, nearly 2 inches. This slime is good stuff: high tensile strength and great stretchiness. They make a net of slime, which they use gladitorially.* They will eat and recycle dried slime; it's expensive stuff to produce and it takes 24 days for them to reload.

There are few groups of animals, and for that matter, few individual species, that don't challenge our human preconceptions. Velvet worms have a social life; groups of up to 15 individuals have been found together. It is presumed that these are related animals. They have been seen hunting cooperatively; some more conservative anthropologists still suggest that cooperative hunting was a human invention. These colonies or family groups tend to be aggressive to other velvet worms that invade their sanctuaries, which are often under a good log. When a kill has been made the dominant female feeds first. They eat woodlice, termites, crickets, cockroaches, millipedes, centipedes, spiders (they will first slime a spider's fangs), worms and even snails.

There is one parthenogenetic† species, but most have a sex life, if not a terribly thrilling one. Females tend to

---

* *Secutor versus retiarius*: swordsman against the man with net and trident; odds three to one in favour of the net-man.
† Going in for virgin birth.

be larger than males, with more legs. The male produces sperm and makes a bundle of it, which he then trowels into the female's genetic opening; some species use a dagger-like or axe-like structure on the head. There are two species that simply slap the sperm-packet onto the female's flanks. From there it is absorbed, though not without causing skin injury. Most velvet worms live at low density. Obscurity is their chosen method in life, and in a quiet and unlooked-for way, it works as well as most other strategies.

Dirty rats

There is no better way of looking hard-nosed than to embrace a misunderstanding of Darwin. Hitler is but one example. People who have reservations about wildlife conservation will tell you that the endangered animals are losers:* weaklings: evolutionary failures. Humans have taken over the planet: deal with it or go extinct. Those that fall by the wayside – pandas, tigers, rhinos – well, sad, but that's life. Or not, of course. Either you learn to adapt to humans or you fail: some writers have taken up the term Anthropocene: a new geological era created by the planet's dominant species. But we humans have always allowed ourselves the luxury of entertaining two contradictory ideas at the same time; in fact, it's the founding principle of human civilisation. So we can despise the animals that fail to adapt to the human domination of the planet – and at the same time, we can despise the animals that have adapted to humans and live alongside us in profusion. Seagulls that haunt landfill sites, pigeons that fly around city centres, cockroaches in our kitchens: all these we treat with contempt. But when it comes to fear and loathing, nothing beats a rat.

The brown rat, *Rattus norvegicus*, is one of the great

---

* It's an interesting philosophical point. The current wave of extinctions is the work of humans, mostly because we have changed conditions – moved the goalposts of evolution, if you like – faster than natural selection can cope. Is it our moral duty to consider non-human animals and amend our ways? Is it in our own best interests to look after the planet we live on? Is it better for our souls, for our sense of well-being, our sense of being at home on this planet, to cherish wild places and the creatures that live in them? Or is it humans first and damn the rest?

success stories of the last millennium. Not known in Britain before the 11th century, it has become a seldom-seen omnipresent fact of city and country life. In a city, you are never more than 3 metres or 10 feet from a rat, so they say: information designed to make our flesh creep. There are more rats than people in Britain, There are supposed to be 100 million rats in New York. We despise rats as scavengers: the more polite and scientific term is commensals, creatures that share the table with humans. But the idea of rats at the table is not overly appealing, so perhaps I'd better drop it.

The muroid or mouse-like rodents have been the triumph of the triumphant rodents, with around 542 species. They are marked by an ability to survive, multiply and adapt quickly to changing circumstances. Rats are the triumph of the muroids. In Graham Greene's novel *Travels with My Aunt*, Mr Visconti, Aunt Augusta's lover of all lovers, war criminal, con-man, thief and so forth, says: "I have a great fellow-feeling for rats. The future of the world lies with the rat... Rats are highly intelligent creatures. If we want to find out anything new about the human body we experiment on rats. Rats are ahead of us indisputably in one respect – they live underground. We only began to live underground during the last war. Rats have understood the dangers of surface life for thousands of years. When the atom bomb falls the rats will survive. What a wonderful empty world it will be for them, though I hope they will be wise enough to stay below. I can imagine them evolving very quickly. I hope they don't repeat our mistake and invent the wheel."

The brown rat is a wet-loving burrower, and over the course of the last millennium, it has mostly displaced the black rat, *Rattus rattus*, which is a climber by nature. The brown rat, spreading from Scandinavia as the scientific name implies (though both species have their ultimate origin in Asia), out-competed the black and sailed to the new world. Rats are a fact of life: an ineluctable aspect of the human condition. I live in the country and keep horses and chickens: naturally, I have rats. They take feed and they pollute feed and they damage stuff, and what's more, they are the cause of ill-feeling among neighbours: so naturally – or unnaturally if you prefer – I control them. Which is a euphemism for killing them: Andy, the rat-man, who used to play darts for Suffolk, baits for them – which is a euphemism for poisons them – and makes sure that numbers don't get out of control.

Control! That's an important part of it, of course. When you read a newspaper scare story related to wildlife, you generally learn that numbers are "out of control". Why is there a need for wild species to be under human control? That's never explained. The idea of losing control, of being unable to regulate the numbers of wild animals we have around us, is something that profoundly distresses us modern humans. It is an insult to our omnipotence, a serious challenge to our understanding of ourselves and our place on the planet. We humans define ourselves by our control – and yet there is a contradiction within 10 feet of you, if you are reading these words in a city. We are not in control. We may be able to extirpate a lot of species with great efficiency, both on purpose and by accident, but there are some species that we simply can't control, no matter how hard we try. They are just too good for us: too smart, too well adapted, too effective. Andy keeps rat numbers down: it's the best he can do. If he ever succeeded in wiping out all my rats, he would only create a vacancy, one that would soon be filled by incoming rats from all over the village. Rats are terrifically effective.

They often carry fleas. The oriental rat flea is the vector for the pathogen *Yersina pestis*, which causes bubonic plague. Rats are not the sole host, but they bear the blame for the pandemics that changed the course of global history, with the Justinian Plague of the 5th century AD, the Black Death of the 14th century and the Great Plague of London in 1665–66. There were 18th-century outbreaks of plague in Marseilles, Cadiz, Messina and Russia. There was a pandemic in China and India from 1855 until, according to the World Health Organisation, 1959. There was plague in Algeria in 2003 and in the Congo three years later. Human history has been shaped by fleas, though we prefer to blame the rats that bore them: you're the dirty rat that killed my civilisation.

Western civilisation has never had many good things to say about rats, for all that they have been roped into thousands of scientific experiments for the very good reason that rats can tell us an awful lot about human beings. But some cultures are less single-minded in their detestation of rats. A rat is the vehicle of the beloved elephant-headed Hindu god Ganesh. I have visited a Jain temple in which the rats were not only fed but honoured. In some Indian cultures, rat meat is considered a delicacy.* In the Chinese horoscope, the year of the rat is by no means inauspicious; its natives are said to be creative, intelligent, generous, honest, ambitious, quick-tempered and wasteful. There is a fine statue commemorating a rat with such qualities in Kyoto; you can find it if you walk the Philosopher's Walk, a stroll that takes you past a number of Zen temples. It is a walk devised for contemplation of higher things than species hatred. What's to hate about rats anyway? Sure, they're ubiquitous, destructive, rapacious and far too effective at breeding. But as you stroll along the Philosopher's Walk, you are entitled to ask yourself: well, who isn't?

---

* The novelist Anthony Burgess claimed more than once (he was immensely keen on recycling) to have eaten rat stew in Chinese-run ships, praising the excellence of their flesh fed from grain they scavenged in the holds, and stressing that their diet made them far more esculent (favourite Burgess word) than rats who lived in sewers.

Another can of worms

When we think of worms, we usually mean the segmented worms, the annelids. They are, to the human mind, *proper* worms, worms that understand what we expect of a worm. We've already looked at flatworms, velvet worms and nematode worms, which are all as separate from each other as they are from humans, all of us in our separate phyla. The segmented worms take the whole thing a step further, and they offer us around 17,000 species in more than 130 families. A number of minor phyla have been downgraded and tidied up into the phylum of annelid worms, as if for administrative convenience, but in fact, it was because of the latest bout of revisionism in the taxonomist's art. The annelid worms include earthworms, ragworms (including those dug up as bait for fishermen), bristleworms and leeches. Have you ever looked out over a river estuary in winter when the tide is low? You will see acre upon acre of soft, fine, beautiful mud, peopled and patrolled by waders beyond counting, birds all prodding their beaks into it. So many of them: and all of them dependent for their lives on annelid worms. If there are that many predators, how many prey animals must there be down there in the wet, salty, fine-grained gloop? Annelid worms live in salt and fresh water and in damp terrestrial habitats. They fill all kinds of habitats in the sea, from the intertidal zones down to hydrothermal vents.

It's worth taking a moment to consider these vents. We used to accept the notion that every food chain on earth begins with the sun. But this bleeding-obvious idea was contradicted by the discovery in the last century of communities that make their living around hydrothermal vents: undersea places fuelled by volcanic energy, like marine versions of geysers, fumaroles and hot springs. These places can support life at up to 100,000 times the density of the surrounding seabed, and they do so without any (direct) reference to the sun. The giant tube worm is a dramatic member of this community. The worm anchors itself with its body and extends a red plume to harbour the bacteria which gain their energy from the minerals that come from these vents. It's an impressive creature, which can be up to 2.4 m, 7 feet 10 inches long.

The segmented nature of annelids gives them movement. By changing the shape of individual segments in sequence, they can move by rippling themselves along. They move by peristalsis: if you like, they sort of *swallow* themselves along. Each segment is more or less the same. The front one is, however, different. Most of us would wildly term this the "head". Quite a good name for it, but it's actually the prostomium: in other words, it's a head, but not as we know it. It contains brains and sense organs, so I'm tempted to ask, what more do you ask of a head? But an awful lot about invertebrate life – correction, an awful lot of all non-human life – is counter-intuitive. The final segment – would humans intuitively call this the bum? – is called the pygidium, and unsurprisingly is the location of the anus.

These more-or-less-identical body segments contain a shared gut, nervous system and circulatory system. Some species breathe through the skin; others through gills. You'd have thought that worminess would be a relatively limited design possibility, but that would be to underestimate the power of evolution, which can outstrip the human imagination any day of the week. The Christmas tree tube worm puts out spiral whorls of blue tentacles; it's a filter feeder from tropical reefs. It's spectacular, in its small way, not much more than 2 inches, 5 cm long. The sea cucumber scale worm has adopted a way of life that parasitises sea cucumbers. The Pacific feather duster worm takes the shape of a daisy. And Gipp's landworm can reach 3 m, 10 feet, in length. Which is pretty impressive, but there are longer worms found in the phylum of ribbon worms. So there's something to look forward to.

Good old Ratty

"Do I contradict myself? Very well, I contradict myself. I am large, I contain multitudes." Words of Walt Whitman. We humans hate rats and mice; we humans are perfectly prepared to love rats and mice. We are all of us at least as large as Walt. I was in Bradfield Wood in Suffolk with people from Suffolk Wildlife Trust. We were there for a project that had touched people's hearts. A sweet, charming, beloved creature has returned to the hazel coppices* within these woods and it has given great joy. The very sight of a single living specimen is enough to trigger the cuteness reflex[†] in all human beings, even the most macho. *Che cariiiino*, how sweeeeet. Nor am I immune, *au contraire*: one look at them and *everybody* goes all soppy. You'd do anything for them. I am reminded of Douglas Adams,[‡] writing on the

---

* Coppicing is an immemorial method of woodland management, and it works particularly well with hazels. The tree is felled to just above ground level; it then produces half a dozen or more dead-straight wands, which were traditionally harvested for building and other purposes. You keep the coppice going by cutting every 10 years or so; the coppice stool spreading and producing more and more wands. A coppice stool can be half a millennium old. The technique produces a dense thicket of hazel which is a good thing for wildlife, especially dormice and nightingales.

† The cuteness response is generally thought of as a useful evolutionary development in humans. Human young need care and protection for many years: it makes sense, then, for adults to feel protective of them. The triggers for this response are large eyes, relative to head, and flat faces. Adults who retain such traits are considered more trustworthy, warmer and more employable than others. Stephen Jay Gould, scientist and writer, described in a famous essay the changes Mickey Mouse has been through since he was created in 1928: his eyes becoming larger, his nose less pointy. In short, he has become cuter. Teddy bears, as already noted earlier, have much flatter muzzles than real bears, and larger eyes: and for the same reason.

‡ Douglas Adams was the author of *The Hitch-Hiker's Guide to the Galaxy*.

near-extinct flightless parrot of New Zealand, the kakapo: "If you look one in his large, round, greeny-brown face, it has a look of serenely innocent incomprehension that makes you want to hug it and tell it that everything will be all right, though you know it probably will not be."

Dormouse. The creature the Mad Hatter and the March Hare tried to stuff into the teapot, who told Alice the story of Elsie, Lacie and Tillie, the three little sisters who lived at the bottom of a treacle well. Linda, the beautiful heroine of Nancy Miltford's *The Pursuit of Love*, is having tea with her awful husband on the terrace of the House of Commons: "'Oh shut-up, Tony,' said Linda, bringing a dormouse out of her pocket, and feeding it with crumbs."

But the dormouse isn't dependent on its literary connections for its charm: charm is its birthright. Tiny, with enormous eyes, and with bright-ginger fur, this is a beast to soften the hardest of hearts. It's a muroid rodent, but we are perfectly prepared to love some of them every bit as much as we hate rats and mice. Suffolk Wildlife Trust found it comparatively easy to raise money from the public and from grants for their dormouse projects: there is nothing controversial or divisive about dormice. They are the sort of creatures we instinctively feel we ought to have around. As a result, they were reintroduced to Bradfield Wood, at first with artificial nesting boxes and artificial feeding, a "soft release" in the jargon, so that they could get the hang of wild living without dying in the attempt. We were there to count them and to see how well they were doing.

Dormice have a reputation for sleepiness. They are serious hibernators, and spend half the year asleep. But when they are awake, they are bold, acrobatic and dynamic. The first dormouse was investigated by Alison Looser, who had just acquired her licence to handle dormice and was working on the project under the wise and experienced Simone Bullion. Alison carefully inserted a hand into the nest box and the rest was a flash of ginger lightning as the dormouse ran up her arm and vanished into the tangled canopy of the hazel. Well, so much for weighing it and gathering other important data: but it looked in pretty good shape to me. (I should add here that Alison's technique is now as swift and certain as you could wish, and she is a crucial member of the Suffolk Dormouse Group.) The rest of the day we progressed rather more calmly from nest box to nest box, counting one dormouse after the next. The dormouse is impossibly tiny in your hand, pushing its face between a tight circle of finger and thumb. Their home is up in the canopy, where they run and leap at high speeds, almost weightless, feeding on hazelnuts in season, flowers, fruit and insects. As I write, the Bradfield dormice are now in their sixth generation following the release in 2006. In 2011 Bradfield held more dormice than any other wood censused in Britain. And joy has been unconfined.

We are prepared to love rats just as much as mice. There is a river – technically a beck, because it's seasonal – that flows through the house I used to live in, with a bridge that crosses it. From this bridge I have seen one of the most entrancing sights of a lifetime spent looking at wild animals: rats. "Absorbed in the new life he was entering upon, intoxicated with the sparkle, the ripple, the scents and the sounds and the sunlight, he trailed a paw in the water and dreamed long waking dreams. The Water Rat, like the good fellow he was, sculled steadily on and forbore to disturb him."

Ratty, good friend to Mole, was, at least to me, the real hero of *The Wind in the Willows*. And out in the Wide World – as the Rat says: "That's something that doesn't matter, either to you or me... Don't ever refer to it again, please" – water voles, as they are more properly called, are met with almost universal delight. Whenever I say that I have sat on my bridge watching two water voles at the same time pattering about in the wet, puddled hollow of the half-dried-out river, splashing, swimming, plunging and squatting on their haunches to nibble pawfuls of salad, I am greeted not with disgust but with envy. I may have rats on my premises: but I also have water voles. Voles are different from rats, but not dramatically. It's the same muroid body plan, the same pointy head and naked tail. One species we love, the other gives us the creeps. David Attenborough confesses to a phobia about rats. But everybody loves dormice and water voles.

Everybody loves barn owls and kestrels as well. I'm not suddenly bursting into the next class of vertebrates and talking about birds here: barn owls and kestrels depend for their survival on muroids, most particularly on the short-tailed field vole. You may have met bank voles if you have a cat and live in a mildly rural area: when the cat brings you a gift of something that looks more like a fat, furry teddy than a slimline mouse or shrew, you've probably got a bank vole. Conservationists used to tell us that barn owls and kestrels are "important" because they keep down "pests" like rats and mice. These days, we'd be more inclined to say that muroids in general and field voles in particular are important because without them we'd have no barn owls or kestrels.

But we are, in certain circumstances, prepared to love rats and mice for their own sake. I have seen dormice dance in the canopy, I have seen Ratty plain. Is that not something to be envied?

Trio for piano, bassoon and earthworm

Charles Darwin was every bit as large as Walt Whitman, but the contradictions he contained were those of method rather than conclusion. When we come to Darwin, we are usually concerned with his ability to think on the grand scale: majestic, sweeping, universal, world-changing. But he was able to reach extraordinary conclusions because his work method was detailed, meticulous and minute. You'd have thought him piffling if you didn't know better. There are big ideas lurking in the tiniest facts – the fall of an apple, for example – and Darwin was at his best pottering about among details of stuff that most people didn't think worth a damn... but always thinking, always theorising, always prepared to make a daring sortie from the great citadel of fact.

Darwin spent many years throughout his life studying earthworms. His last publication, which came out six months before his death, was *The Formation of Vegetable Mould through the Action of Worms: with Observations of Their Habits*. It sold faster than the *Origin* when it first appeared. It was Darwin who established the fact that worms are vital to humans. Plant life on an agricultural scale would not be possible without worms. Earthworms convert organic matter (dead leaves, animal droppings) into humus, in which plants can grow. One wormcast – that is to say, its faeces – contains 40 per cent more humus than a comparatively sized sample from anywhere in the top nine inches, 23 cm, of soil. By ingesting soil particles and passing them out again, worms make minerals and nutrients available to plants. Their constant movement through the soil keeps it aerated and drained. Darwin reckoned the worm population at 53,000 per acre, but that is now considered conservative. Even in poor soil the number is more likely to be 250,000; in rich soil it will be something like 1,750,000; according to other inflationary estimates, six million. In a fecund cattle pasture, the biomass of the worms beneath the surface is likely to be greater than that of the cattle grazing above it. "It may be doubted whether there are many other animals that have played so important a part in the history of the world as have these lowly organised creatures," Darwin wrote.

There are around 6,000 species of earthworms, and they are distributed across the world. They are hermaphrodites; they will come to the surface after rain to meet and mate, it being hard to do so underground. There the couple will exchange sperm and both will go away and produce a cocoon full of eggs that hatch out into wormlets. They are so important to human life that earthworms are farmed commercially. A loss of earthworms would be disastrous for humans – and there is a threat to them in Britain and elsewhere in Europe. New Zealand flatworms have got into the ecosystem through imported contaminated soil and plant pots. They prey on earthworms and have no natural predators in Europe.

Earthworms keep humans alive; they also change landscapes. Darwin examined the work of earthworms around Stonehenge. At his home, Downe House in Kent, he laid out objects on the ground and left them there for 20 or 30 years to see what would happen: how long it would take before the soil, shifted by the worms, consumed them. He dug a trench to see how far they had gone underground. He created a large instrument to measure the action of worms, and called it the wormstone; you can still see it at Downe

House. Darwin was a great worrier at problems. He was always prepared to think in unconventional directions, and was never shy of testing a thought in action. For example, he wondered if earthworms were affected by music. So he got his wife Emma to play the piano to worms, and his daughter Frances tried them with the bassoon. Darwin himself whistled. There was no response: but negative data are also good data.

Many people have wondered about Darwin's delight in such apparently trivial subjects as earthworms. From the meaning of life to the action of worms seems to some a fairly precipitous descent. Stephen Jay Gould, one of the greatest science writers that ever picked up a pen or struck a keyboard (mentioned in the previous chapter's notes on the cuteness response), wrote a spirited defence of Darwin's worms in one of his essays for *Nature* magazine. It was, he said, all about gradualism. Darwin's theory* of natural selection meant that the forces of nature require vast amounts of time for the gradual changes to assert themselves from generation to generation, as antelopes stretched up towards the highest leaves and eventually – eventually – became giraffes, doing so because individuals with longer reach were more likely to live longer and so pass on their advantage to their long-necked offspring – who would do the same thing themselves, and on and on. The action of worms brings about very slow, very gradual changes to a landscape: changes that are imperceptible from day to day, but which are measurable over the course of a human lifetime. And in many lifetimes the changes that worms bring about must, of their nature, be massive: earth-changing. The notion of gradualism could not be demonstrated more

---

* The word "theory" makes all kind of mischief. In loose, conversational use it means a hunch, or a notion: I've got a theory that the stars are God's daisy-chain. But a theory is not a mere hypothesis. In specialised scientific use, a theory is the only possible explanation of the known facts: as in the theory of gravity, and the theory of evolution.

clearly. So there it is: earthworms make modern human life possible and they help to explain how humans – and every other living being – got to be the way they are.

Night-leaper

But we must celebrate the loveliness of rodents once more before we move on, and as we do so – we have no choice – we will celebrate yet again the virtuosity of nature. I was in Namibia. I had been travelling with some people from Save the Rhino, an excellent charity of which I am now a patron. One evening, after a day's hard travelling, we left the blacktop and drove off into a fine wildlife area towards what turned out to be an excellent lodge. It was dark. We got out the bright spotlight and attached it to the car's battery. I explained that I was by far the best qualified to use it. Fact is, spotlighting for game at night is hellishly good sport. The trick is to look for eyes: eyes that reflect back at you in the spotlight, as the eyes of a cat do when caught in a car's headlights. Light bounces back from the tapetum, which is a light-gathering device at the back of the eye. This reflects extra light back onto the retina and helps an animal to see better in low light. We primates don't have them, but most other mammals do.* It was a short drive, half an hour or so; I remember catching a couple of genets from their eyeshine in the trees. But then came the star of the night. I snared it with a glimpse of its eyes and then caught it in the full beam. For a moment it was mine.

And a more improbable thing you'd be hard pressed to

---

* Pigs and squirrels don't, being in theory strictly diurnal; in practice pigs often operate around dawn and dusk, and even at night. Some night birds have a tapetum: owls, most nightjars, stone curlews, and the kakapo, mentioned a couple of chapters back. Why, then, do human eyes sometimes glow red in photographs taken with a flash? The very strong light reflects from the human retina, and takes its colour from the blood vessels that nourish the eye.

find, no matter where you travelled. Somewhere between a hare and a kangaroo with a bit of squirrel and the merest hint of dormouse, the last in the expression of the face, the brightness of the eye (both being nocturnal) and the same ability to inspire the cuteness reflex. But before anyone could say aaaah (*che carino!*), it was gone. It had pinged out of the beam with an abrupt kangaroo leap for which there seemed to have been no preparation, as if it had simply released a hidden spring. I followed it with the spot, leap after leap, dark-tipped tail streaking out behind as a balancing pole. Front paws tucked demurely in front. Then it vanished.

Springhare. I'd seen them before, but this was a specially memorable sighting because I'd done the finding. Springhares do well in areas where the grass is too poor and/or too periodic for antelopes and other large grazers. They thrive on floodplains, when the grass can be briefly superabundant after rains, but at other times is too short and sparse to support big grass-feeders. In other words, this is a vacant niche – or would have been if a rodent, of all things, hadn't adapted to fill it. The springhare is a grass-eating rodent. The requirements of the niche are that the occupant is small enough to find the intermittent grass adequate, yet large enough to have the mobility to reach it whenever and wherever it appears. The springhare can be up to 43 cm, 17 inches, from head to the base of the tail, with another 48 cm, 19 inches, of tail.

But this is a size that brings its own problems. Springhares are small enough to be eaten by snakes and owls and mongeese,* but big enough to be a decent snack for a lion. Springhares have an awful lot of predators to avoid, so they have to be fast: really, seriously fast. The forces of evolution came up with the answer, and a rather unexpected one. Springhare: an unconventional masterpiece of a rodent.

---

* I prefer this plural; mongooses may be strictly correct but it sounds all wrong.

The daughters of Doris

The longest of all animals comes from the phylum of ribbon worms, or Nemertea. This is the bootlace worm and the record is 54 m, 177 feet. This is considered unstretched and a good record, unlike the specimen mentioned on page 89. Even Usain Bolt takes five seconds to travel from one end of *Lineus longissimus* to the other. It's only a few millimetres wide: this is as close as we can get to an animal that has length without width. Euclid called a line "breadthless length" and that's the impression given by the ribbon worms: creatures that apparently exist in a single dimension.

As you will imagine, such a creature is immensely fragile. They're always breaking up, but snapping in half is the last thing to worry a ribbon worm. It just becomes two ribbon worms: or more than two: fragmentation is a matter of routine. They do sex as well; some of them are hermaphrodites and the rest go in for that old-fangled male-female stuff.

Most ribbon worms are small as well as thin: the bootlace worm is exceptional. Most species measure less than 20 cm, 8 inches. It's not a massive phylum: 1,150 species in 41 families, most of them soft, slimy and cylindrical: yet another phylum that shows us how many ways there are of being a worm. Worminess is such a good, straightforward, no-nonsense concept that it has evolved again and again. There are a dozen or so freshwater species of ribbon worm and another dozen that can make a go of it in moist places on land, all of them tropical or subtropical. The rest are marine. Most of these can be found in sediment, or in crevices made by rocks and shells. Some make themselves a semi-permanent burrow lined with mucus. Most creep about, some of them gliding on a trail of slime. A few are free-swimming, moving with an up-and-down motion. They are named Nemertea for the sea nymph Nemertes; she was the daughter of Nereus and, really rather pleasingly, Doris.

Some of them are lovely things, in their worm-like way, coloured yellow, orange, red or green, some with patterns so they look like gently prehensile old school ties drifting through the waters of the ocean. But they are also remarkable and voracious predators, armed with a hidden weapon that operates in the manner of an interplanetary menace from *Doctor Who*. This is the proboscis, a device they keep hidden in a sac on their heads, just above the mouth. When it is time for action, they turn the sac inside out and it becomes a lethal weapon. Some of them have spikes for gripping and/or piercing, and some employ venom to immobilise their prey. The proboscis is controlled by a muscle than can stretch 30 times its own length: a thing of fearsome power. Ribbon worms prey on crustaceans, annelid worms and molluscs, even fish; some scavenge dead animals. They can pick on annelid worms their own size. They're not all fierce, though: a few species filter-feed, and others absorb dissolved nutrients directly from the water through the skin: this is where extreme length becomes a major asset.

All those worms. All those different sorts of worms. And just part of the kingdom of animals which is just one of the kingdoms of life on earth. I remember on that same trip to Namibia, the one on which I caught the springhare in the spotlight, we camped out in the desert one night. We started setting up camp as the sun began to get serious about coming down. As it did so, we had a beer,* obviously. There was a

---

* Namibian beer is rather specially good, something to do with the German influence; it was a German colony from 1884 to 1915. Hansa, Windhoek and Tafel: they all slip down a treat.

star in the sky. There were a good few stars in the sky. There were many stars in the sky. We had another Hansa. There was no moon. There was no light on the ground for many miles in any direction. There was no cloud, it being desert. It had been a still day; there was no dust. And because of all these things, stars followed stars which followed stars: more and more and more and more. The white cloud of the Milky Way looked like a solid thing. Eventually, I was lying in my sleeping bag on my back, wondering at that incredible sky. Had Van Gogh just painted it? No: he wouldn't have dared: portraying a sky like that would have caused him to forfeit all credibility as a painter who worked from life. This was beyond anything he ever painted: there were more white bits in the sky than black bits. And the more you looked the more you saw.

And that's what it's like looking at the Animal Kingdom: more than you ever thought, more than you are capable of thinking of, one after another after another after another. And each in its own way a star.

Flashin' sunshine children

Shrews are not rodents. They were traditionally classified in a catch-all group called insectivores; these days they are found in their own order of Soricomorpha, which contains getting on for 400 species. Like rodents, they look like something the cat brought in, and for the best of reasons. There was a time when shrews were given a hero's role in the history of the earth. They beat the dinosaurs because of their quick wits and nimble bodies, and so paved the way for the Age of Mammals and the rise of good old glorious us. The little shrew-like nibblers created the new and glorious world we inhabit today. This idea is celebrated in a song of Paul Kantner and Jefferson Starship, written by Kantner, Grace Slick and Joey Covington on their album *Blows Against the Empire*. Here's how the little heroes struck their blows against the empire of the dinosaurs:

> Tyrannosaurus rex was destroyed before
> by a furry little ball that crawled along
> the primeval forest floor
> and he stole the eggs of the dinosaur...
> we are the egg-stealers
> flashin' sunshine children
> Diamond thieves...

It's a great image: an outmoded civilisation being destroyed by nimble little utopians in a song for the new age. Alas, the science doesn't stack up. The world was changed not by utopian shrews but by a meteor that struck the earth 65 million years ago. The rise of the mammals was nothing to do with their/our intelligence or their/our long-term planning or anything else to their/our credit. It was a matter of dumb luck.

How to define luck? We are invited to despise the very idea of luck. To say you believe in luck is either a confession of weakness or an elaborate piece of false modesty. But there is a difference between luck and superstition. To believe that you can influence the course of events by crossing your fingers or putting your right cricket pad on before your left cricket pad is pretty silly, even though the process helps some people to cope with uncertainty. But luck can best be defined as the influence on your life of things beyond your control. This taboo subject was explored in a book by Ed Smith with the not inappropriate title of *Luck*. Ed, a friend of mine, is a former cricketer who played three times for England. He was given out incorrectly when looking good in his last innings for England: that was unlucky. Beyond his control, certainly. It's also bad luck when your number fails to come up at roulette: but that doesn't mean it's not your own bloody silly fault if you've staked the mortgage on 17.

A great deal of the history of life comes down to luck. Mammals, and by extension humans, got to be where they are today not because they are better or smarter, but because of the right meteor at the right time. In the same way, an animal might be perfectly adapted to water. What does it do when the climate changes and the place becomes a desert? What does it do when humans drain it? Smith implied that recognising good luck for what it is – not something you have earned, not something you deserve because of some special talent or quality of personality – is a necessary part of living a balanced life. We might think of taking on this concept at the level of species, or at the level of class or even phylum.

Because I am many

And still the stars come out: what can you do but have another beer and gaze in wonder? There are at least 4,000 species of bryozoans out there: a galaxy of them. Most of us don't know that bryozoans even exist, but there they are, living their lives with a desperate urgency to survive, in ways undreamt of by most of us humans, with a baffling ingenuity and in numbers that make us dizzy. And they're all animals, just like us.

Bryozoans are quite ridiculously small, measuring around 0.5 mm, .02 inches. But they don't operate as individuals. They operate as colonies. Many of them look like corals, but they're a good deal more complex. A lot of species form mats on hard surfaces and on seaweeds, and they look a little like moss; Bryozoa means moss animal in Greek. Serious taxonomists tend to prefer Ectoprocta, but the more ancient name has stuck for informal use – informal use, that is, among the few people with whom the subject of moss animals crops up.

They filter-feed by means of a retractable crown of tentacles, beating hairs on them to waft impossibly minute food particles into the mouth, and there's an anus for getting rid of waste: and that's the basic form of an individual in a bryozoan colony. These individuals, all unviable alone, are called zooids. Each one is genetically identical; they reproduce by budding off new zooids. They can also reproduce sexually, but inside the colony they are simply capable of becoming more: I becomes we again and again. The feeding individuals are called autozooids. But colonies are capable of producing non-feeding zooids with specialist functions: as hatcheries for fertilised eggs, as defenders of the colony, even as legs, which allow some forms of bryozoan colonies to creep along. Channels between the individuals of the colony allow the non-feeders to be fed by the feeders. In other words, the individuals of the colony cooperate in the same way that organs in a body do. Most of these colonies are around 10 cm, 4 inches, across but there are some monsters that can reach a metre, more than a yard.

One rum thing about the defensive individuals is that they don't develop until the colony has been threatened. The colony produces defensive zooids in response to a threat: which implies that most individual threats are sublethal in colony terms. All the same, they have plenty of predators: sea slugs, fish, sea urchins, crustaceans and starfish. There are some freshwater species, which get preyed upon by insects, fish and snails. Most are found in tropical waters, and few go deeper than 100 m, 330 feet. But the wild world always loves to confuse us with exceptions and bewildering experiments, and there are some species of bryozoans that prosper in the deep oceanic trenches. Some species grow a calcified outer skin, much as corals do; others are soft-bodied. The moss-like mats gave them their name, but they are also capable of producing more fanciful shapes: bush-like, or fan-like, with a trunk; some like small corals; others like leaves, tufts, and one like an open head of lettuce.

I was on the beach at Flamborough Head in Yorkshire with Anthony Hurd of the Yorkshire Wildlife Trust, and he was showing me the creatures of the intertidal zone. As I walked across a patch of frondy seaweed, he told me that this was horn wrack, and it wasn't a seaweed at all. It was a bryozoan, or rather a bryozoan colony. He described it as "a faunal lawn". It was a deeply disturbing piece of information: there seemed to me at once, at a relatively deep level, to be an important moral difference between walking on a plant and walking on an animal.

That breathtaking breath

It was so close I could smell it. The breath, I mean. The sound filling the ears. How could such a thing breathe like me and you, be like me and you, have a brain like me and you, and presumably be capable, at least to an extent, of thinking like me and you – and yet live here? Here was Knight Inlet in British Columbia, a place where the mountains hit the sea and just keep on going. Right by the shore, the depth was 800 feet. I felt an odd sort of vertigo sitting on top of so much water, so close to the shore and yet so far above anything solid. I was in an open boat maybe 12 feet, less than 4 metres, long with Janie Ray of Cetacealab and Neekas the dog. Janie knows everybody who lives in these vertiginous waters: her work, her study, her joy. So does Neekas, who greets whales with a whale bark never used at any other time. Janie invited me out with her for a day of whale-surveying, and so, clad in clownish clothing, every ounce of which I needed against the chill (clothing which made having a pee a hilarious business, it being a great deal thicker than a half-frozen penis), I journeyed with her across the sound trying to focus binoculars with hands in things like boxing gloves. There was an occasional cry of ravens from the shore, the background burbling of the engine. And then the breath.

Huge. Loud. Visible too, hanging over the water, warm from the lungs and condensing in the vicious cold of the air. Maybe 20 feet off. Humpback whale. I could have dived off the boat and touched him. Shallow dome of the head against the bouncing grey water. How long would I survive in there? But it was home itself for him, my brother, my fellow mammal, my colleague in breathing, my colleague also in the warmth of our blood, the live birthing of our females, in the complexities of our brain, in our deeply social instincts, even in our taste for song. The things that united us were far more than the things that divided us.

The concept of a whale boggles the mind. Life began in the oceans and took its time before colonising the land. These land colonialists eventually produced mammals in an astonishing diversity, as we have seen in these pages: large and small, fierce and peaceful. And then some of them went back to the sea. The seals kept their legs and must give birth on land: they are half-and-halfers, as looked at already. But the whales and dolphins – collectively cetaceans – went all the way. They live in the sea, they give birth in the sea. They only come to the land – sometimes, but quite rarely – to die. A perfectly adapted water beast that can drown: it is a strange and troubling thought. All the same, if you were to examine a skeleton of a blue whale, say, or the comparatively tiny harbour porpoise that you can see off the coast of Britain, you will find a few small useless bits of bone about two-thirds of the way towards the tail. Leg-bones: vestigial fragments; souvenirs of their land-walking past. Their distant ancestors were probably the Raoellidae,* a family of artiodactyls, the group that includes antelopes, sheep and cows. This is all highly counter-intuitive, I know: but the ear is significantly similar – the inner ear, that is, because whales don't go in for external ears, which aren't much help in water.

The whale before our tiny boat – he was maybe as much as four times longer than us – ducked his head and with monumental inevitability the rest of him followed. The oddly small dorsal fin. He slowly poured himself, bit by bit, back into the world below the surface, gathering pace.

---

* This is disputed in some quarters, of course, but that's paleontology for you.

And that glorious, terrifying moment as the tail flukes drippingly soared clear of the water, a massive white-blotched black Y in the sky that briefly stood twice a man's height over our boat and then slithered almost soundlessly below the surface. "Bullet!" Janie sang out, making her ID from what looked like, and probably was, a bullet hole in one of the flukes, a contribution to science from the legion of sportsmen that grace this planet. And Janie took notes of the individual, his behaviour, his position via GPS, just a tiny part of the mass of data she has collected about the humpback whales of Knight Inlet. She has perhaps even more data on the orcas, also known as killer whales, which come to the inlet in the spring.

The variousness of cetaceans is captured in these two species. The humpback is huge, its mouth equipped with great baleen sieves: it ingests a vast multi-gallon mouthful of food, and shoves the seawater back out again through the sieves with the action of its muscular tongue. The orca is smaller – though still big enough in all conscience – and toothed, and is a famous predator. Both species live lives of varying social complexity, which Janie records with detached wonderment, though always with a thoroughly undetached joy behind it. Cetaceans do that to us humans: they lift hearts, they inspire.

They come in about 90 species: 90-odd mammals that decided to act like fish: though fish with a strong mammalian accent that gives them away at every turn. Ishmael in *Moby Dick* concludes that whales are fish and in an informal folk taxonomy they are certainly more like fish than they are like dormice or lions. But most fish swim with a side-to-side motion: cetaceans are uppy-downy, beating the water with their horizontal flukes. They breathe through a blowhole in the top of their heads: the spume of the condensing exhalation, the spout, can be used to identify species of whales and dolphins at a distance. Their vast body size makes them spectacularly efficient at long dives – 30 minutes at a time is not uncommon. But when not plunging deep for food they will stay nearer the surface. Here's a hint: if you are looking for humpbacks, survey the sea before you for seven minutes precisely. If a whale hasn't shown itself in that time, it's probably not there: seven minutes between breaths is easy and comfortable for a humpback.

Getting silly

"Now I do my best to keep things moving along, but I'm not having things getting silly." Words spoken by Graham Chapman in *Monty Python's Flying Circus* when dressed as a full colonel. In that episode he kept popping up in the middle of a sketch to bring it to a premature close, explaining: "Stop! It's getting silly!" Which is rather what I feel about the Animal Kingdom now I've got as far as lampshells. You can call them brachiopods, if you don't think that's silly.

But it is, isn't it? Lampshells are very much like the shellfish you get in your *spaghetti vongole* but with one crucial difference: they're completely different. Your *vongole*, like your oysters and scallops and cockles and mussels and for that matter, your *escargots* as well, are all molluscs. That is their phylum and it contains some of the world's most spectacular and enormous creatures. We'll get to them as soon as we have worked our way through all the silly phyla. Your lampshells look much more like *vongole* than many other molluscs, but they inhabit a quite separate phylum, as different (if we but knew it) from *vongole* as *vongole* are from humans. No! Stop this chapter at once! It's getting silly.

Lampshells are marine animals with twin hard shells. "Who isn't?" the *vongole* ask. And now for something completely the same. Ah, but they're not the same, because the shells – valves, to be more technical, etymologically doors – are on the upper and lower surfaces of the animal, while those of a bivalve (two-doored) mollusc are on the left and right. I know: you could hold up either shell and swivel it though 90 degrees to get the same effect, or so you'd have thought – but that's not what the taxonomists mean. It's a fundamentally different arrangement that doesn't depend on which way up the creature happens to be: it's all in the organisation of the organs of the body.

Lampshells come in two main types, and I'm afraid this is a little on the silly side as well. Like professional athletes, they are divided into the articulate and inarticulate. Articulate lampshells have toothed hinges to the shells and simple musculature for opening and closing them; while the inarticulate ones ("All I had to do was tap it and the ball was in the back of the net") are untoothed and complex. So it's all pretty straightforward, really. Lampshells differ dramatically (well, fairly dramatically) from bivalve molluscs in one way, though: they feed with tentacles, which is a radical departure.

They look less molluscan when standing up; many of them can do so on a stalk called a pedicle, looking faintly plant-like. They are an ancient group, with fossils found as far back as the early Cambrian: these are creatures that go right back to the dawn of multicellular life. There are over 12,000 species of fossil lampshells described, divided into 5,000 genera. The largest fossils measure up to 20 cm, 8 inches; modern ones are around half that at the largest, with most much smaller. Modern lampshells are down to a mere 100 species: survivors of a group that once filled the seas – but they still have their existence and their meaning and their relevance. Life is not silly so far as a lampshell is concerned: and the fittest* still survive.

---

* Fittest doesn't mean strongest. The word "fit" means "suitable", as in a meal fit for a king. It's not the strongest animal that survives; it's the one most suited to the conditions. Being small and weak and good at running away can be a very suitable adaptation, making the animal that possesses these skills much fitter than one that is larger, stronger and fiercer. A human "gets fit" in order to become suitable for, say, running a marathon. This confusion of "fitness" with "strength" is responsible for an awful lot of the misunderstanding of Darwin that takes place among English-speaking people. Mind you, non-English-speaking people have always found plenty of other ways to misunderstand Darwin. The fact is that the truths he revealed are uncomfortable to deal with, no matter what culture you come from: that's why Darwin's revelations about life are so often twisted into some kind of attractive mythology.

Song of the sea

As the humpback whales of British Columbia leave the cold, food-rich waters to travel south, so they give themselves up to song. They will sing on their feeding grounds and they will sing on migration, but it's down in the warm waters off the Pacific coast of Colombia that the humpbacks sing in real earnest, as they sport and flirt and those that find favour mate. Not much eating goes on in these balmy waters: they give their all to the pursuit of each other. Perhaps that's as it should be: at certain stages of your year (or life) you give all your thoughts to work and the issues of making a living; at others you give yourself entirely up to love. You spend half your time working in the cold, the other half seeking love in a warm climate. I have been told that there is one excellent reason for this colossal feat of migration, one not unfamiliar to male humans: male whales can't get an erection in freezing-cold water, so they move south to bask in the hot sexy waters around the equator. Alas, this deeply attractive theory doesn't, as it were, stand up.

But what is certain is that in the warm waters they lift their voices in song. Birdsong has been part of human experience since our ancestors first walked upright on the savannahs of Africa; whalesong has been known for not much more than half a century. It was discovered by an American listening out for Russian submarines off Bermuda; the first recordings of whalesong were made in 1952. The humpbacks were always the champions, though the blue whales of the Indian Ocean are also fine singers. The song of the humpbacks has been described by Philip

Clapham, a leading cetacean scientist, as the most complicated song in the entire Animal Kingdom, with, I assumed, the exception of humans. I asked him for clarification; he responded to this impertinence: "Well, since to date humpback whales have not been observed to produce operas, write choral works or even sing *a cappella*, they don't rival humans in musical ability. The major features of humpback whalesong are its complexity relative to other [non-human] animals, the fact that it changes constantly, and that somehow all the males in a particular population sing the same song and yet keep up with those changes."

The songs have been much analysed and broken down. Four to six units constitute a subphrase which lasts around ten seconds; two of these make a phrase, which is repeated for two to four minutes to become a theme. Themes are put together to become a song, which lasts anywhere from 15 minutes to half an hour. A whale will repeat this song note for note, sometimes for hours, or even days. All whales in any one community or area will sing the same song at any one point in time, though the song will continually change and develop. Each year brings a changed song, significantly developed from the previous year's; analysis over 19 years has shown that the same song is never repeated from season to season. The precise function of the song is not entirely clear, though it is a male thing. It seems to have competitive elements to it, males trying to outsing each other; it seems also to be used to attract females. A group of males has been found singing simultaneously (though not in unison) to a single female. There may also be a territorial function. Is the principal function to repel males or to attract females? Interesting point, though I am inclined to question the actual question here. All love songs celebrate the lover every bit as much as the beloved.

The whales start to sing before they leave Knight Inlet and Janie Ray. The males get increasingly interested in the notion of dominance hierarchy as spring approaches and, with ever-fuller bellies, they can start thinking about migration, warm weather and erections. They will posture to each other – rituals of showing off, or if you prefer, a kind of non-contact combat. And the singing will begin. The steep sides of the inlet and the continental shelf offer something rather special to a humpback: an echo. The song bounces back at them. They seem to be singing to their own echo: Janie puts this down as practice, as honing the voice ready for the meaningful encounters of the warm water. But perhaps it's a kind of duetting, like John Lennon with his love for double-tracked vocals. Perhaps whales simply get lost in the song, enthralled by their own musicality; though the cold water of science must also point out that they may ignore the echo entirely.

There is a continuing mystery in the songs of the whales: but with the mystery a kind of understanding, a feeling of closeness. These great rolling unending love songs sound impossibly remote from us: and yet they are also something we can empathise with: vast echoing symphony-length love songs, changing with the insistent march of culture. There is nothing fanciful in this assessment. It has been demonstrated that the songs of humpback whales from the Pacific Ocean on the eastern coast of Australia completely displaced the songs of the west coast population from the Indian Ocean: "A revolutionary change unprecedented in the animal cultural vocal traditions," the scientists concluded, calling it nothing less than a cultural revolution. A new song, Clapham said, will "spread like a wave across the South Pacific". We humans used to think that culture was unique to us, was what defined us, was what separated us from the rest of the Animal Kingdom. Whalesong tells us otherwise. Humpbacks are our colleagues in breathing: they are also our colleagues in song and our colleagues in the cultural transmission of song.

Dirty beasts

We have celebrated invertebrate taxonomy as a silly joke:
now it is time to celebrate it as a dirty joke. We come now
to the phylum of priapulids, or penis worms. One look at
some of these species is enough to tell you that the people
who named them didn't have particularly dirty minds.
These worms, gentlemen, really do look like your dick.

And – well, that's about it, really. Once we've got the dirty
joke out of the way, there's not much more to say. There are
only 16 known species in the entire phylum, though there
are fossil priapulids from the middle of the Cambrian era;
like the bryozoans, they go right back to the start of multicellular life. Back then, it seems, they were a great deal more
numerous. They are not vastly relevant to modern ecosystems, but we have to give them credit for hanging on. They
mostly like marine sediment, and some of them can live
several kilometres deep. They range from (readers are invited
to make their own size-matters joke at this point) from 0.5
to 20 cm, 0.2 to 8 inches, in length. The small ones feed on
bacteria, the larger on small slow-moving invertebrates.

Some species have an extensible spiny proboscis which
gives the surreal dick-like effect. They come in male and
female forms and they mate in the male-female manner
we can relate to. And various characteristics make it abun-
dantly clear that they are not closely connected with any
other form of life on earth. They aren't afterthoughts; if
anything, they are beforethoughts: a phylum of not-quite-
obsolete animals that still make a living in the silt of obscurity, remote from human consciousness.

6. PRIAPULID

Gnomes of the river

Perhaps cetaceans represent the most extreme form of mammal: the land animal that went back to the sea, and in doing so, produced the blue whale, the largest animal ever to make a living on planet earth. You'd have thought that this was enough, somehow. But the forces of evolution came up with an extreme form of the extreme form: and what's more, they did it not once but six times.

Here's an important thing to learn if you go chasing wild-life: if you base a trip around a search for one particular species, you will often be disappointed – but you might see something even more marvellous while you're looking. And anyway, if you go to wild places there's always something. I tried, then, not to be too disappointed when I failed to see tigers in Nepal. These things happen, and anyway, the forest was pretty wonderful. I then moved on to Koshi Tappu for the waterbirds. One afternoon we made a trip to the Koshi Barrage:* a vast water management device strung across the Koshi River, which eventually joins the Ganges. And there I encountered one of the weirdest things I have ever seen.

Well, three of them, to be precise. An unearthly pallor about them. A ridiculous shape: like dolphins drawn by a small child. A long beak and round forehead. Each one about 6 feet long. And leaping from the water, as if it were suddenly boiling. They were Ganges river dolphins, dolphins who have abandoned the oceans to live in the murk and

---

* The man-made embankment by the Koshi Barrage gave way in 2008, causing disastrous flooding which made a million people homeless and affected 2.3 million in Bihar in India.

silt at the bottom of rivers in the Ganges system; there is a subspecies that does the same thing in the Indus. They are almost blind; they seek out fish by echolocation, and with the dextrous use of that beak, which is four times longer than the beaks of any of the species of marine dolphins. River dolphins are supposed to be much less active than the marine ones, but the Ganges dolphins like a bit of a leap every now and then. "In turbulent water, when disturbed by a boat, and sometimes for no apparent reason at all, Ganges river dolphin leap," says the frankly baffled Lyall Watson* in *Whales of the World*.† And here before me were three of them, leaping for no apparent reason at all. They reminded me of a dolphin version of the gnomes that live deep beneath the ground in CS Lewis's *The Silver Chair*: "All carried three-pronged spears in their hands and all were dreadfully pale. Apart from that, they were very different; some had tails and others not, some wore great beards and others had very round, smooth faces, big as pumpkins." The parallel isn't precise, I know, but they are all pale and they all look bizarre and they all live in deep dark depths, leading a life that we surface-dwellers find dark, dismaying and unwelcoming.

There are six species of river dolphins, or rather there were. There's the Ganges and Indus dolphins, which make up one good species between them, sometimes called the South Asian river dolphin. There is the Amazon river dolphin or boto, and the River Plate dolphin that inhabits salty estuaries at the southern tip of South America, the Araguaian river dolphin and the Bolivian river dolphin. There was also, until very recently, the Yangtze river dolphin. These

---

* The same Lyall Watson that gave us *Supernature*, an attempted scientific study of the paranormal, a major hit of the 1970s. Watson also did television commentary on sumo wrestling.

† When I first looked at the pictures of Ganges river dolphins in this book, I assumed that the illustration was the work of an incompetent draughtsman; obviously no living dolphin could look like that. It is, of course, accurate in every way, just as the pictures of the aardvark were.

species are not closely related: rather, the same basic idea has evolved four times over.

The Yangtze river dolphin, or baiji, was declared functionally extinct in 2006 after a 45-day expedition found no trace of a living specimen. Since then, an individual has been videoed, but alas, any surviving baiji belongs to a class known informally among scientists and conservationists as the Living Dead: members of a species that has no future. Overfishing and damming of the Yangtze has helped to do for the baiji, along with sound pollution. Any one has been swimming when there is a motorboat in the water will understand: water carries sound unnervingly well, and the noise of an engine fills your ears even when it's a couple of hundred yards away. The busy highway of the modern Yangtze is forever filled with the sound of engines: how can a dolphin echolocate in such circumstances? How can it even think? Douglas Adams, hunting for the baiji in *Last Chance to See* in the 1980s, pondered the life of the dolphin in the modern Yangtze and suggested that it was like a blind man trying to live in a disco. "Since man invented the engine the baiji's river world must have become a complete nightmare."

Us alone

It is in our nature to set humans apart from the rest of the Animal Kingdom. In an ideal taxonomy, if we could ignore such irritating things as facts, we would classify ourselves as a single genus, family, order, class. Better still, a single phylum: we ourselves us: us alone; *sinn fein*, which means we ourselves in Gaelic: we happy few, proudly and eternally alone. But we are cousins in apehood with chimpanzees and bonobos, we are colleagues in singing and breathing with the great whales, and we have an unbreakable bone-deep relationship with snakes and frogs and with everything that we loosely refer to as fish. There are, to our eternal dismay, thousands of us.

If we want a single species that is also its own phylum, we must forget our proud selves and meet *Trichoplax adhaerans*: undisputed master of a phylum which includes no other. The placazoans are gloriously and disturbingly singular: one is one and all alone. This is the phylum that doesn't do diversity. Almost uniquely, the lone placazoan has its being in uniqueness.

Well, you can argue a second species if you like. *Treptoplax repians* was described in 1896, but nobody has seen it since, so it is probably kinder to forget it as a brief hallucination rather than explain it as a creature good to go. *T. adhaerans* rules the placazoans: a ruler without rival.

The name means flat animal, which is fair enough; flat is what they are. Blobs, really. Flat blobs, up to 3 mm across. They have never been seen in their native habitat, no one knows that they eat in the wild, and it's not even clear if they have sex. Or to put it another way, we don't know much, do we? But they are something to do with the shores of warm oceans. They were discovered on the glass walls of an aquarium in a laboratory.

They are wonderfully simple, with only four different kinds of cells. Even sponges have at least ten, some as many as 20. We mammals – all of us, not just humans – have more than 200. Placazoans are good creatures to argue about: it has been suggested that they could be the oldest branch of multicellular life. There is another theory that they are descended from more complex animals but have simplified and prospered, in their quiet way, as a result. That goes against the conventional myth of evolution: that it is a process of working towards ever-greater complexity, ever closer to perfection. But if simplicity works, evolution, not being bothered by any goal save that of survival, is perfectly prepared to drop both its dignity and it complexity and live.

We humans like to think that we are loners, but we're really not. We share our houses and our lives with dogs and cats and goldfish. We have meaningful relationships with horses:* that happens in cultures right across the world. We respond to birds because like birds, we are creatures of sight and sound: we share musicality with birds as well as with humpback whales. Before we had records and hi-fis and iPods, we had caged songbirds to brighten our homes and our days with music. Our sense of continuity with our fellow vertebrates makes the world more meaningful and more comfortable. We are, when we think about it, not only one of many, but happy to be so. It is not us but the placazoan who is always and implacably alone.

---

* Many books have been written about a relationship between a human and a horse; I've written one myself: *The Horsey Life*, same publisher.

Disgustingly upside down

Mammals can exist deep in the ocean and sing their hearts out, so it's hardly a big deal to find them taking to the air and navigating by radar as well. A quarter of all mammal species are bats. They range from whoppers with a wing-span as wide as a man is tall to ridiculous little things you could slip into your waistcoat pocket.

True flight. That's the thing about bats. We've met some of the gliders and swoopers. I've always loved watching flying squirrels make their reckless and apparently doomed journeys across forest clearings, but for all their ingenuity and dexterity the general direction of each journey is down. Gliding is really just falling with attitude. But bats fly under their own power. When a bat is up it stays up.

You see them in England as flickering shadows caught in the tail of your eye: flying things that don't move quite like other flying things. Partly it's because they are so agile, especially laterally: you see them jink sideways, with no effort at all, to catch an insect, seizing it in a leathery embrace of the wings and whisking it mouthwards. I remember a bat moment in Brisbane, where I was covering the cricket (English bats not faring too well that day) for *The Times*, part of the sportswriting side of my life. There is a small relict patch of mangrove swamp in the Brisbane River, so naturally, I took a twice-daily detour through there on my way to and from the ground. One evening, I left the cricket-ground later than usual. It was getting dark by the time I stepped onto the boardwalk. As if at a signal, the crickets struck up in unison, a high, mad, electrophonic squeal like an alien weapon sending the whole world crazy in a science-fiction film. At more or less the same instant huge shapes appeared in the sky and powered towards me, strong and majestic, eventually flying directly overhead: vast, arched, measured wingbeats: big fruit bats on their way to a night's plunder. They gave the city a richness it hadn't had before.

And a slightly sinister touch, of course. We humans find all the creatures of the dark a little difficult, and give them all kinds of frightening attributes. Owls are birds of ill omen in Europe; birds of death, in Africa. Nightjars are surrounded by strange names and curious legends. Bats terrify the life out of people. The association of bats with vampirism goes very deep, even though there are 900-plus species of bats that have nothing to do with blood, and those that do are pretty small. I suspect it's something to do with the naked skin of the bat's wing: so unbirdlike. People from bat groups who try and help people with bat roosts in their outbuildings or (scarier still) in their attics will all tell you the same thing. They will catch a bat and show it to the householders, who see the furriness of the body and the bright mammalian eyes. Oh, it's just like a little mouse! And with this revelation the bat becomes tolerable, even agreeable. Strange, in a way: we humans go in for naked skin rather than furriness: but the naked skin of the bat repels while the furriness makes it acceptable. DH Lawrence wrote a poem about it; here are the closing lines:

> Wings like bits of umbrella.
> Bats!
> Creatures that hang themselves up like an old rag, to sleep;
> And disgustingly upside down.
> Hanging upside down like rows of disgusting old rags
> And grinning in their sleep.
> Bats!
> In China the bat is symbol for happiness.
> Not for me!

A bat is defined by flight. Flight is not a bonus: flight is all and everything to a bat. They come in two suborders, which might loosely be described as little buggers and bloody great big buggers. The big 'uns, the Megachiroptera or flying foxes, comprise a single family. They can weigh up to 3 pounds, 1.3 kilos, and look very impressive, especially when flying across the Brisbane River. Of these, only one genus uses echolocation, and a fairly crude version of it; the rest rely on their excellent sight.

The Microchiroptera contains everything else: proper bats, you might say, echolocating and flying. The two things are very closely linked. Both are expensive in terms of energy, but when echolocation is used in flight, synchronised with the wingbeats, it becomes far more economical. In short, for a bat in flight echolocation becomes a freebie. This is the breakthrough that has allowed bats to prosper in so many places across the world: within the Arctic Circle and over almost every bit of land on earth. They are only absent in Antarctica and a few very remote islands. In New Zealand, the Azores and Hawaii, they are the only native mammals; that's to say, the only mammals not brought there by humans. This has led to such unexpected creatures as the New Zealand short-tailed bat, which drops to the ground and scurries about like a mouse wearing a cape, ground-feeding in the manner of shrew.

In most circumstances a bat will let out a single echolocating squeak with every wingbeat: up to 15 every second. The problem of echolocation is the same as the problem with radar: you have to find a way of not hearing your own sound. A bat has to shout and listen at the same time, and this is accomplished by means of a muscle in the inner ear that acts like an earplug, switching itself on and off so that the bat hears only the echoes, which it uses to build a sonic world-picture in its brain. Bats are also able to compute out the Doppler shift: the phenomenon that distorts sound when it is mixed with high speed (the change in note you can hear from a passing racing car).

Bats are capable of high energy and very low energy, and they inhabit the extremes of the temperature spectrum as a result. They can reach 41 degrees Celcius in flight; they can go torpid and drop to 2 degrees at rest. Flight and echolocation mean that the night is theirs: and therefore, they have the great food resource of night-flying insects almost to themselves. Flight is expensive in energy terms, but distance becomes relatively cheap: it is harder to fly for an hour than to run for an hour, but it is easier to fly a mile than to run a mile. That is the conundrum that gives bats the range they need to exploit this resource. But insects aren't the only food. Some bats take fish, frogs, other bats. Many (not just the flying foxes) take fruit, also nectar. As a result, some trees and many cacti are bat-pollinated, a concept that disturbs our sense of what is right.

But not half as much as the thought of bats that drink blood.

Lacing Venus's girdle

Obscure and heartbreakingly beautiful, the Ctenophora propel themselves through the world's oceans on tiny beating hairs called cilia (getting cilia, in other words). Their name means comb-bearer; they are sometimes called comb jellies, sometimes sea gooseberries. Their beauty comes from their colour: most species are bioluminescent; that is to say the produce their own light and they glow. But there is a still lovelier thing that some of them do: as they waft themselves through the seas with their cilia, they can produce shifting rainbows that run up and down their bodies; ask YouTube to find you Ctenophora. This is not colour they produce themselves; it is a trick of the light, or if you prefer a more grown-up explanation, it is a phenomenon of physics called diffraction.

The Ctenophora make up a phylum of their own. Quite a lot of them are quite a lot like cnidarians, the corals, sea anemones and jellyfish we have already met in these pages, but the differences are marked enough to separate them. Like the cnidarians, their body is a mass of jelly, with a layer of cells outside and another as a lining. But the outer layer is two cells deep in ctenophores, rather than the one in cnidarians. They operate in a similar way: both need water to flow through the body cavity for digestion and respiration, and they have a decentralised network of nerves rather than that brain thing some of us have.[*] Some scientists used to put the two in the same phylum, but the

---

[*] What is described by Woody Allen in *The Sleeper* as "my second favourite organ".

current trend is to consider them separately.

They are predators. Hard to believe it: they look like harmless drifting patches of loveliness. Unlike cnidarians, they don't have any stinging cells, but they possess sticky cells that they use to catch prey, often at the end of tentacles. Microscopic larvae and small crustaceans are their main diet and they can eat ten times their own body weight in a day: which seems like a lot for a lightshow to eat. There are between 100 and 150 species described.

If they're not cnidarians, what are they? Some have seized on evidence, most notably the anal pores, and suggested that this indicates a strong affinity with animals with bilateral symmetry (as opposed to the radial symmetry of a jellyfish). Bilaterans include insects, worms, snails and all us vertebrates. There are molecular data that contradict this, but not strongly enough to settle the argument.

Most Ctenophora are free-swimming, though there is one group that creeps along the bottom. They occupy all parts of all oceans: from poles to tropics, from surface to benthic depths, from inshore waters to the open ocean. Most are pretty small, a few millimetres in length, but out in the open ocean there are some whoppers, a metre and more across. They are so fragile that you can't catch them in a net: they just fall apart. One species is an absurdly lovely thing called Venus's girdle: more like a piece of abstract elegance designed to recall a wing than a living creature. Fossil Ctenophora have been found in Devonian rocks, and fossils very *like* Ctenophora go right back to the Cambrian.

I believe I once swam with a swarm of Ctenophora, snorkelling off a beach in Barbados when playing truant from an assignment to cover the cricket. They were mostly transparent, so much so that they almost weren't there at all. There was a sweetness and innocence about them that I mistrusted entirely: not knowing what I was dealing with, I flipped my fins and changed direction. I think now that this was a bit windy of me: human deaths by sea gooseberry are very rare.

The altruistic vampire

If humans were more like vampire bats the world would be a much happier place. It's altruism I'm talking about here: the free giving of something; something you need for yourself but are prepared to surrender to another. We humans like to think that altruism in any form is uniquely human: a real and above all moral division between us and the rest of the Animal Kingdom. Vampire bats contradict this view.

I really do mean altruism. Giving to blood relations doesn't count. It isn't altruism to surrender things for your children: that's just self-interest at the genetic level. The old notion that charity begins at home is a rejection of charity. It's only when you give – free, gratis and for nothing – to an unrelated individual that you can claim genuine altruism. And it is the bedrock of the society of vampire bats.

Vampire bats do at least drink blood. All kinds of fanciful legends have built up around them, but that part of it is not made up. There are three species; the white-winged vampire and the hairy-legged vampire prefer the blood of birds, and are adept at clambering through the branches to reach nests and nestlings. Common vampires – we'll call them plain vampires from here on – prefer the blood of mammals: mostly livestock in these modern times, horses and cattle; their wild prey is tapirs, deer, peccaries and agoutis. They will on occasions drink human blood. The bite is not painful (apparently), but there is a problem in that 0.5 per cent of vampires carry paralytic rabies.* The bats can also contract the disease, and populations sometimes crash as a result of infection. Vampires are pretty small: 9 cm or 4 inches long, a wingspan of 18 cm, 7 inches. They are adept on the ground, crawling with agile speed to reach the target and search out a convenient blood vessel. They use razor-sharp incisors to remove the skin; the resulting wound has been described as looking like a golfer's divot.

The anti-coagulant in their saliva allows the blood to flow, and they lap up the trickling blood like a pussycat with a saucer of milk. They weigh about 40 grams; they can take a meal of half their own body weight.

This sounds as if all the odds favour the bats, but not so. It's quite common for a vampire to go through a night without a meal. This is inconvenient, to say the least: a vampire can starve to death in three days, and a hunger-weakened bat obviously finds it harder to find food. Blood-hunting is a skill that improves over time: yearling bats will fail one night in three; more experienced animals will only fail one night in ten. But all the same, failure is a fact of life for even the best blood-hunter.

Vampires are social creatures, coming back to the same day-roost at the end of every night. There are some big colonies, up to 2,000 animals, but most are much smaller, and centre on a core population of females. Some of them will be related, others not: females often change their day-roost at some stage in their lives. All members of a roost, then, related or not, are likely to know each other well. And so it happens that a bat who has failed to find blood in a night's flying will beg a meal from a neighbour. From a friend, we would say, if we weren't so terrified of sounding

---

* Gerald Durrell, forgetting this fact, tried very hard to get bitten when he travelled in South America, sleeping with his foot sticking out from his bedding. He was anxious for the experience, in the interest of science and as ever, in search of a good story. He remembered about the rabies after a couple of frustratingly unbitten nights.

anthropomorphic. The friend will then regurgitate blood, thus sharing a meal. Under this system females have been known to live for 15 years.

Reciprocal altruism is still altruism; obviously this is a system that works on a mutual back-scratching basis. Rival journalists at a football match, heads down and writing hard, often miss the goal and the goal-scorer. "Christ, who scored?" is a cry of pain that goes up almost every time the ball hits the net. We who saw it never keep the information to ourselves: after all, it'll be us next time. Human society – humane society – depends on the small kindnesses that you perform as a matter of course and that you expect to be performed for you in turn. But it's not just humans who are humane.

Here be mud dragons

The smartest thing mud dragons ever did was come up with a retractable head. They use it for getting about: unusual use for a head but they are unusual animals. First they unretract the head by filling it with body fluid. Then they anchor it: it is profusely equipped with spines called scalids. Once anchored, they retract the head again: and behold, the dragon has dragged itself forward. This is the phylum of Kinorhyncha, which means moveable snout. The name mud dragons makes them sound rather more glamorous than they really are: tiny scraps of life, the biggest no more than a millimetre in length, living in mud and sand at the bottom of the sea and feeding on the even tinier scraps of organic matter they find there. Their limbless bodies are divided into 13 little chunks called zonites: you can argue, should you wish to, about whether or not these are true segments.

Their bodies are covered with a tough cuticle which they moult a few times as they grow to adulthood. Some have simple ocelli – basic eyes – on their heads. There are 180 species described, and they can be found in marine mud from the intertidal zones to depths of 8,000 m, and from Greenland to Antarctica. They are simple and effective creatures. They didn't have the potential or perhaps more likely, they just lacked the breaks needed for great complexity and huge variety. As we have seen, luck plays a massive part in the way life operates. The dinosaurs had bad luck 65 million years ago when that meteor struck the earth: for mammals, and for one particular species of mammal, it was the break they needed. Luck is more important than innate superiority. Mud dragons don't look like conquering the world, though you can never tell; no one would have bet on us mammals either. Besides, in the lottery of life it is honour enough to survive.

Pocket dynamo

My experience of marsupials has been limited to what I've managed to pick up in bars. I once experienced two species together in a bar in Brisbane. The first was the thylacine, or Tasmanian tiger. It's extinct now, alas, but you can still see it on the label of the excellent Cascade beer of Tasmania. The second was a possum. The waiter went to the garden to throw out some peanuts: a possum descended the tree, took them from his hand and accepted strokes. Naturally, I had to join in. A half-grown possum was riding jockey-style on its mother's back: it had no interest in peanuts and was unwilling to be stroked, but the mother was very agreeable.

Which shows something of how tenacious and how successful marsupials can be: hanging on as urban bandits despite the brutal concreting of their environment. We tend to see marsupials as freaks and aberrations: perhaps that was why I found the idea of marsupials so thrilling at my junior school. I remember doing a project about Australia and its marsupials precisely because they seemed, in so many ways, to be so remote from Streatham, where I grew up. There are getting on for 350 species of marsupials currently recognised, and 70 per cent of them can be found in Australia and New Guinea. The rest are in the Americas, mostly South America. There are 13 in Central America and one, the Virginia or common opossum, spreads as far north as the United States.

We humans are placental mammals: all us placentals give birth to young in varying degrees of good-to-go-ness. We do a good deal of growing and developing inside the womb.

Marsupials give birth to blobs which they raise outside the womb: in the famous marsupial pouch. A baby koala weighs one-fiftieth of an ounce: a bare scrap of almost nothing.

These slug-like, utterly unmammalian-looking things are a little repulsive, in both the appearance and the idea. Sure, it's only what we placental mammals do inside the womb, but doing it outside seems indecent. Partly as a result of this, we are inclined to be just a little patronising about marsupials, looking on them as exotic primitives, mammals that tried to reach the dizzy heights set by us placentals but fell woefully short. As is often the case, these stock human responses to the natural world tell us more about our own nature – our desire for both separateness and superiority – than they do about the animals themselves. A female kangaroo is able to cope with three young at the same time, all in different stages of development; not many placentals can claim that, least of all those the size of a kangaroo. Humans are exceptional in their ability to do so.

There are marsupial equivalents of many of the placental mammals. The thylacine on my beer bottle was as cool a carnivore as any that you'd find among the official family of Carnivora. It went extinct because it got a taste for sheep when the white settlers introduced them, and was shot out.[*] There are gliding marsupials, as we have seen, burrowing marsupials who live as moles, and ant-eating marsupials. There is an extinct marsupial not unlike a rhinoceros, and another kangaroo-like marsupial with a neck like a giraffe. Another species was considered a marsupial lion, though

---

[*] There are still occasional claims of thylacine sightings in Tasmania, where they hung on after they had gone extinct on the mainland. If there are any left, they are surely, like the baiji, functionally extinct in the wild: members of the living dead. But what if humans could intervene and start a captive breeding population and reverse the extinction? This is what happened with the Arabian oryx. They were extinct in the wild; the world population was down to seven animals brought together from zoos across the world to Phoenix Zoo. They are now back out there in the wild as viable, though heavily protected and guarded, animals. There's a similar story about the European bison. It is a pleasant fantasy to think that this might yet be done for the thylacine: though fantasy is all it is, in all probability.

some people now argue that it was a tree-climber and used its impressive teeth to eat fruit.

We regard antelopes and other even-toed ungulates as supreme processors of grass: the kangaroo is every bit as good. Australia moved northwards as a result of tectonic action, and as it did so, the thick forests were replaced by more open habitats. To exploit such places you need to be able to process grass and cope with the exposure to predators that comes from being out in the open all the time. Antelopes and other placental grazers have teeth that grow continuously throughout their lives, allowing them to deal with a diet that is very demanding and wearing of teeth. Kangaroos can't do that, but they operate a different system that works just as well. As their front teeth wear out, so the teeth behind them migrate forward and engage. It's a method that allows a red kangaroo to live and graze for 15 to 20 years.

Kangaroos are a family, not a species. The superfamily of kangaroos, wallabies and rat-kangaroos comprises 68 living species, and they're called macropods. The red kangaroo, the giant, the one we think of when we hear the word kangaroo, is the most recent: it's only been around for the past five to 15 million years. Far from being primitive, it's the very latest thing, in evolutionary terms. And it hops.

Why did they develop as hoppers rather than gallopers? It looks like a mistake, a fudge, a compromise, but it's astonishingly efficient. Perhaps it's the ideal gait for a pouch-bearing animal: better to carry a pouch full of baby vertically than horizontally if you are looking for sheer uninhibited speed. Hopping is pretty ungainly at low speeds, but once they move beyond 15 kph, 9 mph, kangaroos start to stretch out and shift. The hop is much more energy-efficient than a run because with each bound energy is stored in the tendons of the bent hind legs. In hopping you get a certain percentage of your forward motion as a freebie: like riding a pogo-stick. The reds can cruise at a sustained speed of 35 mph, 55 kph, and can touch on 60 mph, nearly 100 kph, in bursts. And they can clear a 3 m or 10 foot fence.

The Hamlet worm

They're the suicide worms. Reverse Samaritans. They make their living from the suicide of others: an unusual strategy. They come from the phylum of Nematomorpha, a confusing name that means they look a bit like Nematode worms even though they're nothing to do with them. The members of this phylum live a four-part life: as an egg, as a pre-parasite, as a fully-baked parasite, and as an adult. The species *Spinochordodes tellini* parasitises grasshoppers, crickets and locusts. This invasion gives the animal an infection that affects its brain, and thus its behaviour. The unfortunate grasshopper is compelled to seek out water and drown... thereby allowing the parasite to leave its host and return to the water, where it becomes an adult. It can then go wormily squirming off to find another adult and make eggs. This behaviour has been the subject of experiments: the degree to which the infected grasshopper actively seeks water is uncertain, but certainly its behaviour becomes very erratic and it tends to end up drowned. In short, the creature is driven first to madness and then to suicide by drowning. What Hamlet did to Ophelia, this nematomorph does to grasshoppers.

Nematomorphs are sometimes known as horsehair worms. They are often found in water troughs, and it was once thought that falling horsehair somehow turned into worms. They are also referred to as Gordian worms, a much better name, because they are prone to knotting. Some of them measure 50–100 cm, 20–39 inches, and you even get the odd whopper up to 2 m. They are never thicker than 3 mm, 0.12 inches. Most of them are found in streams and puddles, water troughs and cisterns. There are a few species that make a go of it in damp earth and five that have moved out to sea. The adults of these seagoing worms are to be found in plankton; the larvae parasitise crabs. There are about 350 species of nematomorphs described; most of them parasitise beetles as well the grasshopper group, the orthopterans.

Their game plan is to do their eating as parasites and then take to the water to reproduce. They are staggeringly simple: the adults have no excretory, respiratory or circulatory system. But they are adept at survival: they can, for example, survive even after their host has been eaten, wriggling out of the predator to seek another victim. Some species employ an interim host, technically called a paratenic host. They don't affect this host: they just will it to die. When the host is consumed they will parasitise the consumer.

Death comes for the Elephant's Child

I was overwhelmed by grief and it was the grief of an elephant. I can see her now: pacing hopelessly backwards and forwards in the long Groucho-Marx stride that elephants adopt for extremes of emotion, and I can hear her, too. For she was howling with grief. Her behaviour was so startling, so odd, it took a while to understand what was going on. It was night. There was a group of elephants around Mfuwe Lagoon in the Luangwa Valley, plainly visible in the spotlight. You must use a spotlight sparingly with elephants: sensitive animals, they find it intrusive and disturbing, so you must point the light away from them and observe them in indirect light: a Rembrandt chiaroscuro of mysterious and compelling shadows. The small herd consisted of half a dozen females and as many young ones. They were all milling around in a dreadfully troubled way: and one of them, distress in every line of her body as she moved desperately up and down, was screaming. It was a sound that terrified; it was a sound that broke my heart even before I understood.

But understanding came and it was terrible. The elephants had been drinking in the lagoon, some of them wading; there were black tidemarks on their grey pillared legs. Elephants, as the world knows, love water. So do crocodiles. And there was disturbance in the lagoon as well as on land: ripples and splashes and an awareness of big things moving about, big things happening. And then I saw it: in the lagoon the pathetic floating body of a drowned elephant. Not much more than a baby: a small child, if you like, a year old, just about. Crocodiles don't move very much; when they do they strike like lightning from a clear sky. Here was Kipling's comic story of "The Elephant's Child" retold as tragedy. The sound of elephantine grief made it clear that no lesser term would do. Kipling's elephant was saved from the crocodile by the enigmatic Bi-Coloured-Python-Rock-Snake, but not before its nose had been stretched into a trunk. Not this time.

It is not anthropomorphising to say that the elephant was in grief. It is the plain obvious truth. She had witnessed the death – sudden, cruel and horribly dramatic – of her offspring. Of her son, or her daughter, of her child. And she was in a state of enormous distress. She clearly knew and understood what she had seen, and equally obviously, she didn't take it lightly. Oddly enough, we often acknowledge the notion that grief is not restricted to humans when talking of some dreadful incident: people will say that "she made an animal groan" or "an inhuman cry escaped his lips". Equally, when people write of dreadful incidents in the wild world, they sometimes say an animal made "an almost human sound".

It is clear that grief can be observed in animals other than ourselves. And naturally we would expect to witness it in elephants. Elephants, whales and dolphins and the apes are all remarkable for their intelligence: and intelligence seems to involve recognisable emotion as well as the ability to do things like communicate, make plans and solve problems. Elephants have always been important to humans. If we are embarrassed to have an affinity with monkeys, we are proud to share at least some concerns with elephants. In Graham Greene's *Travels with My Aunt*, Aunt Augusta produces words she had copied years before from the writings of St Frances de Sales,[*] and asks Henry to read it to her.

---

[*] St Frances de Sales, 1567–1622, was bishop of Geneva and had an anachronistically gentle view about divisions in Christianity. He wrote *Introduction to the Devout Life*, which, again surprisingly for the time, was aimed at laypeople.

"'The elephant,' I read, 'is only a huge animal, but he is the most worthy of beasts that live on the earth, and the most intelligent. I will give you an example of his excellence; he...' The writing ran along a crease and I couldn't read it, but my aunt chimed gently in. 'He never changes his mate and he tenderly loves the one of his choice. Go on, dear.'

"'With whom,' I read, 'nevertheless he mates but every third year, and then for five days only, and so secretly that he has never been seen to do so... But he is seen again on the sixth day, on which day, before doing anything else, he goes straight to some river wherein he bathes his whole body, for he has no desire to return to the herd until he has purified himself.'"

This is not very accurate as field observation, but it is very helpful indeed in telling us about the warm feelings humans have for elephants. I wrote an account of the shooting of an elephant in my novel *Rogue Lion Safaris*. I didn't make it up. The incident was accurately – more or less slavishly – based on an incident involving my old friend Chris Breen, who was then working as a guide in the Luangwa Valley. He was so traumatised that he gave up guiding and returned to England. Years later, walking in the bush, he and I came across the bones of that elephant: he was moved beyond speech. Later, he told me the story in immense, second-by-second detail. He had put himself into extreme danger to protect his clients. The shot that killed the elephant almost certainly saved his life. I put every detail into the novel. It's my words, mostly; it's Chris's voice, mostly:

"'Why did you shoot him? Why did you shoot him?'

"Tears rivered unregarded down my face. It was at this point that I was seized by the urge to grab Aubrey's gun and to batter him senseless with the butt. I could have done easily, so hotly did the adrenaline run. I think Aubrey knew that, for there was fear in his face. But I turned away, dropped into a crouch and sobbed unashamed."

Grief can work both ways where elephants are concerned.

Cynthia Moss is the great matriarch of ethology, the study of animal behaviour. Her work on elephants, showing the ties of relationship and affection that animate elephantine society, has changed the way we look at elephants, at the wild world. She, too, writes of grief. Here are some lines from her wonderful *Elephant Memories*, telling of the events that followed the death of an elephant she called Tina at the hands of ivory poachers.

"Teresia and Trista became frantic and knelt down and tried to lift her up. They worked their tusks under her back and under her head. At one point they succeeded in lifting her into a sitting position but the body flopped back down. Her family tried everything to rouse her, kicking and tusking her, and Tallulah even collected a trunkful of grass and tried to stuff it into her mouth. Finally Teresia got around behind her again, knelt down and worked her tusks under her shoulder and then straining with all her strength, she began to lift her. When she got to a standing position with the full weight of Tina's head and front quarters on her tusks, there was a sharp cracking sound and Teresia dropped the carcass as her right tusk fell to the ground. She had broken it a few inches from the lip well into the nerve cavity, and a jagged bit of ivory and the bloody pulp was all that remained.

"They gave up then but did not leave. They stood around Tina's carcass, touching it gently with their trunks and feet. Because it was rocky ground and the ground was wet, there was no loose dirt: but they tried to dig into it with their feet and trunks and when they managed to get a little earth up they sprinkled it over the body. Trista, Tia, and some of the others went off and broke branches from the surrounding low bushes and brought them back and placed them on the carcass. They remained very alert to the sounds around them and kept smelling to the west, but they would not leave Tina. By nightfall they had nearly buried her with branches and earth. Then they stood vigil over her for most of the night and only as dawn was approaching did they reluctantly walk away, heading back towards the safety of the park. Teresia was the last to leave. The others had crossed to the ridge and stopped and rumbled gently. Teresia stood facing them with her back to her daughter. She reached behind her and gently felt the carcass with her hind foot repeatedly. The others rumbled again and very slowly, touching the tip of her trunk to her broken tusk, Teresia moved off to join them."

Grief. Loss, mourning and the understanding of death: these things are not unique to us humans. We are not alone in facing the terrible things that happen in the course of a life: nor are we alone in understanding them.

Who needs oxygen?

This book would have been shorter if I had written it in 1982, instead of working on a rightly unpublished novel and seeing if I could make beer come out of my ears. It's not that my stamina has improved: they discovered a new phylum in 1983. That is to say, a group of animals as different from anything else on earth as we humans are from butterflies: and one no one knew about it – which goes to show that there is an awful lot going on on this planet. They are Loricifera and they live in the spaces between gravel. Not big, then: they range from minute to microscopic. Under magnification they look rather like a vase of flowers: a tentacled container. The container is called the lorica, hence the name. They are found in the sea at all depths and latitudes. They have a head, mouth and digestive system, but no circulation. There are about 100 known species, of which about half have been described and named. They are both abundant and ubiquitous, though it took us a long time to find out about them.

They sound a pretty undistinguished lot on the whole – but the group contains creatures that can claim to be the strangest animals on the planet. Three species of loriciferans live at the bottom of the l'Atalante basin in the Mediterranean. Not only is it totally dark down there: there is no oxygen either. The water is so salty that it is near saturated and doesn't mix with the layers above. That is why there is no oxygen: but the loriciferans thrive down there nonetheless. Some living things from outside the Animal Kingdom – bacteria and viruses – thrive without oxygen, but this is an animal: that is to say, a multicelled being from the same division of nature as ourselves. Loriciferans are the only animals of that degree of complexity that can live without oxygen. Unlike the rest of us, these three loriciferans don't operate at the cellular level on mitochondria, which require oxygen. Instead, they employ organelles driven by hydrogen.

This is so peculiar that it forces us to change our definition of life, and with it, our idea of the conditions necessary for life to take place. Multicellular animal life can exist without light and without oxygen. That is quite interesting in a philosophical sort of way. It is quite startling when you start thinking about life beyond the earth. Our search for life on other planets is no longer restricted to oxygen-rich environments. Once again, we find that life is weirder than we think: and more varied than we think. And perhaps more widespread than we think.

Epiphany

Charles Darwin and James Joyce are the twin heroes of this book, and it's high time we had a bit more Joyce. Joyce was very keen on epiphanies. Epiphany is the feast of January 6 or Twelfth Night. It commemorates the Three Wise Men and their visit to the stable in Bethlehem. Epiphany means a showing-forth: in this case, the showing-forth of God. The young Joyce collected epiphanies: significant moments when something was shown forth. Here he is, speaking through his alter ego Stephen Dedalus, in the third chapter of *Ulysses*: "Remember your epiphanies on green oval leaves, deeply deep, copies to be sent if you died to all the great libraries of the world, including Alexandria? Someone was to read them after a few thousand years, a mahman-vantara. Pico della Mirandola like. Ay, very like a whale. When one reads those strange pages of one long gone one feels that one is at one with one who once…"

So here is another wildlife epiphany of mine, deeply deep, naturally, and it's here to mark the moment when the vertebrate cycle of this book makes its first significant shift. We're at the end of mammals. Apologies to all those who didn't make the cut; all those untold epiphanies I might have related: the day's march to find rhino, the African wild dog den, the beaver in Canada, the gathering of 60 hares in Suffolk, the otter in front of my house. Not all of my stories have made the book, and not every mammal either: but never mind. I haven't even told you about the dolphins of Cardigan Bay and the Moray Firth, ay, very like a whale. But it will soon be time for the birds, for if we tarry much longer with the mammals we will never unravel the riddle of when a fish is not a fish.

"Ladybird, ladybird, what is your wish?

Your wish is not granted unless it's a fish."

As the Incredible String Band inimitably expressed it.

I was in Africa; Zimbabwe this time. Walking in the bush is a balancing act: you want to get really quite close to really quite dangerous animals without putting yourself in real danger. So we walked quite close to quite a lot of elephants. As we did so, we found rather a lot more elephants. And in doing so, found even more elephants. In fact, we were completely surrounded. A circle, untidy but unquestionably of 360 degrees, of elephants, the nearest maybe 50 yards off. The bush here was pretty thick, which explains why so many elephants had sneaked up on us; or rather, how we had inadvertently sneaked up on them. Not the remotest idea of how many, couldn't really see, but well into three figures. There was just me, my wife Cindy and our guide Graeme Lemon. Graeme dropped into a squat: "We'll just rest up for a moment," he said. So we also got down on our haunches. And put ourselves at the convenience of elephants.

All around us, the elephants were involved in a swishing, crackling, mutual encounter: this was a meeting of some magnitude, with much social interaction going on. We had made ourselves small and insignificant, unthreatening. The elephants were interested in each other, not us. So we waited. We waited until they should be pleased to let us go. We could, I suppose, have broken the circle, blasted our way out of it, Graeme firing his gun to the sky as we did so. It would have been dangerous, but perfectly possible. But instead, a moment of respect. Thus we humans deferred to the elephants.

That's why the moment, the long moment – we must have squatted there for 20 minutes – has stayed with me. It was a drastic reordering of the usual priorities. For once, we humans counted second: second to elephants in massiveness, in dignity, in importance, in urgency. Our convenience was secondary to that of the wild world. We were in a somewhat vulnerable position, but instead of making drastic adjustments, the wise thing was to accept this secondary status.

All around us the bush snapped, crackled and popped, and the elephants, mostly hidden, raised dust that drifted towards us and stung our nostrils like pepper. We could smell the farmyardy smell of elephant dung and hear the excitement in the air as they went through their social obligations. Slowly, they drifted away, as if the wind had caught them, dispersing, some of them aware of us, but giving us a wide berth, wary but unthreatened. And after a long period of silence, we stood again, flexing sore knees, and made our way back to camp. Knowing that something had been shown forth: some important thing about the wild world and our relationship with it. It was a moment of confirmation: one that spelt out for all time the nature of my relationship with the wild world.

Deeply deep.

A bit samey

Everywhere you look in the Animal Kingdom it's the same story: the more a design gets successful, the more it improvises around a theme. Evolution is the eternal jazz musician, making it up as it goes along, taking an idea and seeing where it will go, and doing so again and again and again, never sated. The radiation of antelopes shows us how a basic plant-eating design can encompass creatures as different as a giraffe and a royal antelope, which is not much bigger than a hare. And we haven't even got to insects yet: still less to the beetles that contribute so many mind-boggling numbers to insect variety. There are beetles everywhere: and everywhere they diversify, adapt, form new species. In their impossible variety, beetles tell us the basic story of the way life on earth works: making more and more and more different kinds of living things. The chaetognaths tell another tale: perverse, contradictory, perplexing. There are many of them; in some places they are the most numerous life form – and yet they obstinately refuse to do variety. Many individuals, very few species: that's the upside-down way in which the phylum of chaetognaths runs its affairs.

They're sometimes called arrow worms: shaped a bit like darts or torpedoes, which makes them adept at moving through the sea. Most of them are part of the great oceanic soup called plankton, and they make their living by eating other forms of plankton. They are armed with spines, which they can cover with a hood to aid with the streamlining. They sense prey from its vibrations and subdue it with neurotoxins. They are transparent, and pretty small, 2–120 mm, .08–4.5 inches long. And there are only about 120 species of them, even though they are uncountably numerous when it comes to individuals. Most are free-swimming and planktonic, undulating along in the vertical plane in the manner of whales, even though they are somewhat smaller. Others attach to the substrate, or to algae; some drop down to benthic depths. The surface-dwellers tend to go for a daily vertical migration: dropping down during daylight to avoid predators and rising to the surface at night to feed. Their sometimes colossal numbers give them a considerable ecological significance: predators themselves, they are much preyed upon by fish. For all their obscurity, they are an important part of the way the ocean sustains itself, and it seems that they always have been. Fossil chaetognaths have been found as far back as the Cambrian: they were there as multicellular life began, and they're still hard at it. They are all hermaphrodites, swapping sperm to fertilise each other's eggs. Their courtship is a cautious business: the aim is to establish intimacy without edibility, a conundrum that exists in many areas of invertebrate life. They are both few and many: a multitudinous contradiction, yet another group that shows that nature doesn't necessarily keep to its own rules.

Time for transition

There is a slightly apologetic tone to *The Origin of Species* – and behind it, a more than slightly defiant one. It's obvious that if Darwin's theory of evolution by means of natural selection is true, there must be transitional forms: inbetweenies; missing links. The problem for Darwin was that he didn't have any to hand. Trust me, Darwin said. I know I'm right. Bear with me and the half-and-halfers will surely make themselves known in the fullness of time. With the air of a man who knows he is trying the patience of his audience, he points out the paucity of the fossil record. Animals are only fossilised in unusual circumstances, so the record is necessarily patchy and incomplete. You can't expect to find the fossils you want: it's a miracle that we have any at all. Naturally, this very obvious weakness was seized on by the many opponents of the book. Ha! they said. Anyone can pluck a theory out of the air. He's just made it up as he went along. It's all just speculation: and jolly bad speculation at that.

It looked like the answer to a prayer. It looked like divine intervention: as if God had stepped in on the side of the materialists. Because in 1861, two years after the publication of the *Origin*, the greatest of all transition fossils turned up in a quarry near Solnhofen in Germany. It was obviously a bird. Equally obviously it was a reptile. It was of course archaeopteryx. It had the bony jaws and tail of a reptile, it had teeth like a reptile and it had claws on its fore-limbs. But it also had feathers: beautifully, unmistakably printed in the joyfully receiving slabs of German lithographic limestone.

If it had feathers it had to be a bird: no other possibility existed.* The clawed fore-limbs were wings. But it wasn't like any bird alive today.† It was a link between one order and another: indisputable demonstration of the truth that birds evolved from reptiles.

There was another non-missing link, if only Darwin had been sure of it. He knew about the duck-billed platypus; he had even, bizarrely, attended the slaying of one. When the *Beagle* called in on Australia, Darwin went for a platypus hunt, and his host killed one. "I consider it a great feat to be in at the death of so wonderful an animal," he wrote. The duck-billed platypus had created a scientific sensation when the first skin came to London in 1798. Many considered it a hoax: "A high frolic practised on the scientific community by some colonial prankster." If you examine this historical specimen, still kept at the Natural History Museum, you can apparently still see the scissor marks around the bill, where a scientist named George Shaw tried to find the stitches he was certain attached bill to pelt. But the platypus's bill is by no means the oddest thing about it. The females lay eggs. And, unusually among mammals‡ (but not reptiles) the male platypus is venomous. It carries poisonous spurs on its hind legs: not lethal to humans but capable of causing considerable and lasting pain. The egg-laying business was only known for certain in 1884, too late for the *Origin*, but a nice incidental piece of confirmation nonetheless. A poisonous, egg-laying mammal breaks all the rules§ – apart from those

---

* It has now been established that feathers first developed among dinosaurs and some now argue that archaeopteryx was a feathered dinosaur rather than a true bird. Whatever: it's still a transition species of the most thrilling kind.

† A South American bird, the hoatzin, hatches out youngsters with claws on their wings, and they clamber through the branches much as archaeopteryx must have done. They lose the claws as they reach adulthood and learn to fly.

‡ Some shrews have venomous saliva.

§ Platypuses (the plural is problematic: platypi is incorrect because the word comes from Greek rather than Latin, platypodes is correct but rather pedantic) also have a remarkable electrical sense: they are able to find prey by means of electrolocation. Creationists will tell you that this proves Darwin was wrong: you can't be primitive enough to lay eggs, and yet advanced enough to use electricity. Why

that Darwin laid down. It is a perfect transition animal.

Come. It is time for this book to make a transition. We have dwelt among our fellow mammals for too long: lost in the furry byways and the milky pathways of our own kind. Mammals can trace their ancestry through reptilian lines; a different fork of the same trunk produced birds. Let us now turn to the feather-bearers: the most studied animals on earth. Birds are visible and audible: their main senses are the ones we humans use. Most mammals understand the world through their noses, but birds flaunt themselves before our eyes and our ears. Get ready to sing, get ready to fly, get ready for the birds.

on earth not? We humans are advanced enough to build spaceships yet primitive enough to possess vestigial tails. We have already noted the modernity of kangaroos, with their leaping gait and their shifting teeth, despite the ancientness – if you like, the primitiveness – of the marsupial lifestyle. Evolution doesn't upgrade just for the sake of it. If something works, it works. So stick with it, until there's a good reason not to.

The peanut trick

They may be just another bloody worm to you or me, but the Sipunculids don't see it that way. What other worm can turn itself into a peanut? What other worm has its arse on its back? These are not negligible considerations for the phylum of Sipuncula, sometimes known as peanut worms. And if anyone gets sniffy and tells you these worms should be considered with the annelid worms like the earthworm, feel free to say: "Oh yeah? What about the complete lack of bristles then? And how do you account for the absence of segmentation?" That'll fix them.

A sipunculid does the peanut trick in response to threat: it retracts its body until it resembles a peanut kernel. That's about as exciting as it gets for a peanut worm: though the odd position of the anus in the anterodorsal region of the trunk is a clear mark of distinction. The phylum is, it has to be accepted, one of the duller ones, so it makes perfect sense that its most famous species is named for the game of golf. The first sipunculid was described in 1827 by Henri Marie Durcrotay de Blainville, and he called it *Sipunculus vulgaris*. A further peanut worm was later described by E Ray Lankester, who named it for a golfing holiday he had taken in St Andrews, and for the professor who gave him the specimen (or possibly for the garment he played golf in): *Golfingia macintoshi*. He renamed de Blainville's worm *Golfingia vulgaris* (as if golf could ever be considered vulgar) and the phylum became Sipuncula.

They're worms, yes. Mostly pretty small, 4 inches, 10 cm is typical, though the range is 2–720 mm, 0.08–28 inches.

They are marine, mostly found in shallow water. They have an unsegmented trunk, and a retractable section which is confusingly known as the introvert. The introvert has a mouth surrounded by tentacles. They have a digestive system and a nervous system, and also a cerebral ganglion or nerve ring that acts as a brain. They can reproduce sexually or asexually, doing so by fission. They mostly live by burrowing; some of them can bore into solid rock, while others live in abandoned shells, in the manner of hermit crabs. Estimates of the number of species vary, the top estimate being around the 300 mark.

A jelly made from sipunculid worms is a considered a delicacy in Xiamen, which is in the province of Fujian in China.

Feather

Birds fly.

I've never quite got over it. Let's start with a kestrel:

> Dapple-dawn-drawn Falcon, in his riding
> Of the rolling level underneath him steady air, and striding
> High there, how he rung upon the rein of a wimpling wing
> In his ecstasy! then off, off forth on swing,
> As a skate's heel sweeps smooth on a bow-bend: the hurl and gliding
> Rebuffed the big wind.

The best description of a flying bird ever penned,[*] and with Gerard Manley Hopkins, my heart in hiding stirred for a bird.

And it's all about feathers. The heart and soul of a bird is summed up in a feather: a feather defines what a bird is, and is the essence of what a bird does. Which is fly. A feather is the perfect implement for flight: immensely strong and immensely light, stiff yet perfectly flexible. You couldn't make or imagine a better device for powered or gliding flight. How amazing that the forces of evolution came up with this wonderful thing: you want flight, here's a thing called feather.

But like a disturbing number of things in life – both wild life and your own life – it was a complete accident. Feathers did not evolve for flight at all. Flight was just a

---

[*] "The Windhover" by Gerard Manley Hopkins, subtitled "To Christ our Lord". Windhover is an old name for kestrel; another, perhaps wisely not used by Hopkins, is wind-fucker.

lucky bonus. We have 10,000-plus species of birds on this planet, living their own lives and illuminating human ones, and almost all of them doing both by means of their ability to fly – and it all came about as a bi-product of the fact that dinosaurs felt the cold. Some species of dinosaurs evolved feathers for thermoregulation: an example being *Dilong paradoxus*, related to *Tyrannosaurus rex* but living 60-odd million years earlier. They weren't fliers, they weren't even half-and-halfers attempting to fly. For them, feathers were nothing to do with flight. They were theropod dinosaurs and their experiments in insulation filled the air with birds.

How did feathers make the transition from overcoats to wings? It's Just So Story time, and you can take your pick. Either they found feathers increasingly useful when jumping up – getting higher and travelling further as the arrangement of feathers improved – or when gliding down. Either way, feathers led to true flight – as opposed to falling with attitude – and true flight led to the conquest of the air.

Birds are descended from dinosaurs, then, and some people prefer to say that birds *are* dinosaurs, feathered reptiles as adept at exploiting the modern world as their ancestors were during their 100-million-year heyday before the meteor strike. But if they are feathered reptiles, they're not much like reptiles as we normally understand the term. Flight changed everything. As they evolved from the first rudimentary jumpers or gliders, the whole flight business grew increasingly sophisticated. Reptilian arms became wings, the breastbone became a keel, to which the big muscles for flapping are attached. The body is powered by a mighty four-chambered heart. Birds have a breathing system that is far more efficient than the in-out method we mammals use: one in which hollow bones and air sacs are constantly used to flush out stale air. Flight is highly expensive in terms of energy: the birds' breathing system gives them a constant supply of the freshest possible air. A human marathon runner would kill for such an advantage, and having found it, would be unbeatable. Some birds, most obviously the skylark, use this system to sing for minutes at a time without seeming to take a breath; it can do so because it is taking in new air all the time.

A warm-blooded, well-fuelled, high-energy body has allowed a bird's brain to develop far more than that of a reptile: and as a result, birds have adopted a vast series of complex and difficult lifestyles. They invest time and energy into the raising of young: this is at the heart of what both they and mammals do, and what other vertebrates mostly don't. It is all based on a high level of intelligence: some birds are as smart as the great apes.

So, as we complete our transition into the world of the birds, let us remember that they are not placed second because they are in any way secondary to us mammals... but then, if you have read so far, you may well be already prepared to abandon the ideas put about by mammal chauvinist pigs. Birds do not belong in the class of mammals, but that's not because we're upper class and they're not. The class of Aves is full of flight: and that is something we humans have always both rejoiced in and envied.

We dream of flight. When we imagine superior beings, we give them flight: Superman can fly, and so can angels. At the church of the Holy Trinity at Blythburgh in Suffolk, the angels that vault the roof float on the wings of marsh harriers. All the non-confrontational sports are about flight and the defiance of gravity: skiing feels like flying, horses give you wings (for every rider, every horse is Pegasus). Gymnastics is the closest a human body can get to flying under its own power. And even golf, contested by people who don't even break out of a walk, is about flight: every golfer who ever struck as ball will tell you it's not about winning and it's not about putting the ball in the hole – it's about making that ball fly. It's about bird-envy.

Emily Dickinson called hope "the thing with feathers".[*] The Holy Spirit came down in the form of a dove.[†]

On the first page of *Goldfinger*: "Yes, it has certainly been time for him to die; but when Bond killed him, less than twenty-four hours before, life had gone out of the body so quickly, so utterly, that Bond had almost seen it come out of his mouth as it does, in the shape of a bird, in Haitian primitives." Birds carry a colossal weight of human mythologies and human meanings on their lightweight hollow-boned frames. We have always thought ourselves a class above our fellow-mammals: but when we seek to express the notion that we are superior to the brutes and the dumb beasts, we turn our eyes to the sky and find it in the birds: in aspiration, in flight, in hope, and in feathers.

---

[*] Woody Allen, being a professional pessimist, entitled a collection of his writing *Without Feathers*.

[†] This point is taken up in the blasphemous song "The Ballad of Joking Jesus", sung by Malachi Mulligan in the first chapter of *Ulysses*. It is adapted from a poem written by Mulligan's prototype, Oliver St John Gogarty.
"I'm the queerest young fellow that ever you heard.
My mother's a Jew, my father's a bird."

No sex please, we're bdelloids

Some species of bdelloid rotifers haven't had sex for several million years, but we'll get to them in a moment. Let us first consider the miraculous appearance of the classic rotifers: the ones that gave the phylum its name, which means wheel-bearers. They look as if they are carrying a rapidly turning wheel on their heads (insofar as it can be termed a head). This is a bigger shock to the perceptions than it sounds: it is not possible for organic life to contrive a mechanism that spins like a wheel, independent of the rest of the structure. The sight of it makes you feel slightly dizzy, faintly seasick. It looks like a piece of clockwork: perhaps the kind used to power Dali's watches. The first instinct is to say that no, they simply can't be animals; they can't be part of organic life as we know it. They look like science-fiction hybrids: part animal, part mechanism: a soft machine.*

A wheel that spins independently of the rest of the structure is a human invention. It required immense imagination to leap away from the forms found in nature: that's why its invention was such a radical step.

The rotifer wheel is an illusion. It is not a wheel but a mouth surrounded by a corona of cilia: tiny beating hairs that move in a coordinated and rhythmic fashion. This coordination is what makes it look as if the thing is going round and round, much as warning lights on a level crossing look like one light jumping from side to side, rather than two

---

* *The Soft Machine* is a novel by William Burroughs, and as such, a description of the human body.

lights going on and off rhythmically. The corona is used for feeding, and in some species it also drives the animal along. This is another example of something that evolved for one purpose successfully used for a second purpose; like feathers in the previous chapter.

Most of us feel as if we may have heard of rotifers. They crop up in pond-dipping experiences and in aquariums. They are microscopic, or near microscopic – there are one or two real monsters measuring 2 mm, .08 inches. They are mostly freshwater creatures, though there are a few that live in salt water, and some that can operate in moist soil and on mosses and algae. Some are free-swimming, some anchor themselves, some crawl along the bottom. There are 25 colonial species. All in all there are (estimates as usual vary) between 1,800 and 2,200 species. They form an important part of freshwater life. Many of them are part of freshwater plankton and are an essential part of the food chain; most are consumers of organic detritus, and so play a big role in keeping fresh water habitable and life-sustaining. It is this talent as cleaner-uppers that makes them useful in aquariums.

We have known about them since 1696. They possess both a mouth and an anus, which makes them incontestably animals. They even have a small brain. And they're odd about sex. Many of the bdelloid group reproduce by means of parthenogenesis, or virgin birth. A bdelloid rotifer can make more, genetically identical bdelloid rotifers, apparently just by thinking about it. And as said above, it is believed they've done so for many millions years: finding no need for genetic diversity as they spin their wheels and live out their aqueous lives.

Not all rotifers take this austere view of sex. Other species are pretty keen on it. Usually, the female is bigger; in some cases as much as ten times bigger. Some species

produce degenerate* males that don't even possess a digestive system. They are, quite literally, only interested in one
thing, and are equipped with a penis to fulfil the one function they have been born for.

---

* The word is used here in its scientific sense to mean something that has passed
from a complex to a more simple state.

The nausea of
Charles Darwin

Birds are beautiful. There's no ducking it. They're not incidentally beautiful: they're beautiful for no reason other than to gladden the senses – and that's a tough nut for a Darwinian to crack. The hard-eyed functionalism and bracing reductiveness of Darwinian thought seems to come a cropper when you look at birds whose beauty apparently possesses no other function than its own loveliness. Birds really are bright and beautiful. "The sight of a feather in a peacock's tail, whenever I gaze at it, makes me feel sick," said Darwin.

There is beauty to be found among us mammals, that's not in dispute. But the beauty of mammals tends to be incidental. I have never seen anything quite as beautiful as a night-hunting leopard, strolling as if every joint had been bathed in a gallon of oil and glowing as if lit from within. But this beauty is functional: the glorious feline gait tells us about the speed and skill of the hunter; the immaculately maculated coat enables him to hide, both from his prey and from those (*che brutto*!) that might do him harm. A leopard is beautiful in the manner of an athlete in competition at a track and field meet. Many birds are beautiful in the manner of a female athlete scrubbed up, made up and wearing her posh frock. A female athlete in competition is beautiful quite inadvertently; a female athlete togged up for some big occasion is beautiful on purpose. A female athlete in such circumstances is beautiful like a bird.

You can start with the orange bill and sleek black plumage of the male blackbird in your garden (the female is much less conspicuous) and go on to the plumes of the birds of paradise, the ridiculous elongated tails of the paradise flycatcher, and then to the ultimate swaggering debating point of the peacock, erecting his tail and shivering all over so that the many eyes in the great fan fluoresce and strobe in the sunlight.

There is no reason for a peacock's tail other than its beauty and splendour. It certainly doesn't help the bird to live longer: *au contraire*, it's a serious handicap, one it sheds rapidly when it's no longer required. I once startled a peacock that was hiding in a single isolated large bush. I heard something inside the bush and walked around it half a dozen times trying to catch sight of whatever it was, expecting a small bird, thrush-size at best. A peacock with a full tail suddenly lost his nerve at my incessant circling and exploded out of the bush with a great scattering of twigs, alarming the hell out of me as he did so: it was like a hundred pheasants getting up at your feet all at the same time (male pheasants being another example of gratuitous beauty). It was a revelation (epiphany) of the cumbersome nature of the tail: it seemed to take about ten minutes to go past. It needs a huge output of energy to get the whole damn thing up in the air; which is no doubt why the peacock first trusted to concealment rather than escape. The shocking suddenness is part of its withdrawal tactics; if you can make a predator jump, you've won yourself a crucial half-second advantage. Life with a tail is hard for a peacock; but it's no fun at all without one.

Darwin cracked the conundrum eventually: that was his way. He took every problem and ground it relentlessly in the rollers of his mind until he had turned it into powder. He came up with the breakthrough notion of sexual selection: something that "depends not on a struggle for existence but a struggle between the males for possession of the females; the result is not death to the unsuccessful

competitor but few or no offspring". That's the end of the line for his genes, as we now understand the situation. In other words, males compete for females, and those who please the females most (and who also succeed in repelling male rivals most effectively) are those who get to pass on their genes. So if females like, say, a red breast, then the bird with the reddest breast will find favour while a bird with a duller breast won't. It makes sense, then, to flaunt that breast, both at females and at rival males: which is of course exactly what the robin in your garden does.*
Competition is at the heart of Darwinism. Different species compete with each other as to which is most suitable – most fit – to exploit a certain ecological niche; individuals of the same species also compete. And competing for the finest female available is one aspect of this competition.

And here's a snag that made the idea of sexual selection very unpopular when Darwin first introduced it: the implication of sexual selection is that the casting vote on the future of a species lies with the females. Males come up with suggestions; females make the decisions. The traditional Victorian world-view didn't find it easy to accept that sexual selection – that female choice – was an important force in shaping the world we live in, and was perhaps a factor in determining who and what we humans are today.

Sexual selection says how: and that's how a peacock's tail came about. It doesn't say why. Why have peafowls chosen, over the centuries, these extravagances of blue and green? What aesthetic preferences guided them to these choices that led to this maddest of avian affectations? How come that again and again in the class of birds, very clear but very different aesthetic standards were established?† Why do birds desire beauty?

---

* The great ornithologist Chris Mead claimed to have observed a robin displaying at his own glorious red beard.
† A question explored in David Rothenberg's remarkable *The Survival of the Beautiful*.

Evolution in reverse?

Life starts simple and gets more and more complex. That's the idea; that's the whole bloody point, isn't it? You start off with slime mould and end up with the blue whale. And us, of course. The whole idea of evolution is to make ever-more complex creatures: a great unfolding saga of improvement. With us as the crown of creation.

And then you get to the Acanthocephala, or thorny-headed worms. They have taken the reverse direction. They have simplified. They have lost the gut, and most other organ systems have been drastically reduced – apart from those to do with sex. They don't even have a mouth. They were originally considered a phylum of their own; now they are more frequently regarded as greatly modified rotifers. Greatly simplified rotifers: they have discarded complexities, all the better to live the life of a parasite. Once established in the gut of some vertebrate, they just hang there absorbing nutrients. And that's where the thorny head comes in handy: all the better to attach themselves to the gut wall. The proboscis has a collection of spines or hooks and the whole thing can be withdrawn into the body cavity. Once established in the gut, there the thorny-heads dangle: I have seen a picture of the gut of an eider duck that is lined, positively furred, with these pendant worms. There is a record of 750 worms in a single duck. The loss of nutrients to these parasites is a problem for the host, of course, but the greater problem is the ulceration of the gut lining caused by the thorny-heads and their uncompromising trick of attachment.

They have abandoned complexity in body plan, but they have certainly acquired complexity in life cycle. Their basic plan is to exploit two hosts: as a juvenile, this is usually a crustacean or an insect, typically a beetle or grasshopper. But they need this host to be eaten by a vertebrate if they are to become adults. One species tends to parasitise a crab that lives in darkness and obscurity, but once infected by the thorny-head in question, it abandons the ways of its kind and becomes an urgent seeker after light, sometimes climbing clear of the water onto exposed rocks or emergent plants: more or less begging to be devoured. It's a pattern not unlike the Hamlet worm in the chapter on nematomorphs, but it gets even cleverer.

Once the first host has been devoured the worm can become an adult and set up residence in the gut of the devouring vertebrate. These hosts are usually fish, though all classes of vertebrate are options. Inside the host they meet, have sex and produce eggs: and the eggs are passed out of the host with the faeces… which is in turn devoured by some scavenging crustacean or insect, which then becomes host number one for the newly hatched juvenile. There are more than 1,000 species described, most of them less than 20 cm or 8 inches long, though some measure up to 60 cm or a couple of feet.

So it's not survival of the cleverest, the most beautiful, the most complex: it's survival of the fittest. The most suitable. And the great simplifications of the thorny-heads make them superbly suited to the task of parasitising the guts of vertebrates. This is not evolution working backwards, for with evolution there is no forwards and no backwards. Only survival. Here is a good word: teleology. It is about purposeful direction: in *Ulysses*, Mr Deasy says: "All history moves towards one great goal, the manifestation of God." Teleology is about being goal-orientated: but evolution is not teleological. Evolution is not driven towards a great goal of complexity. Evolution is just about getting through the day.

How many ways of catching a fish?

It looked so exactly, so perfectly as it was supposed to that I could hardly believe it was real. This was my first sight of a toco toucan: the one* with the orange and black bill and the black and white plumage: the classic toucan from the Guinness advert. It was as if I had been taken back to the Garden of Eden, the one you can read about in the book of Guinnesses.† How gloriously absurd: is this the beak of beaks? I wondered that the bird didn't tip nose down and crash out of the trees into the Paraguay River. You can't believe, watching with mammalian eyes, that such a thing could be airworthy.

But the forces of evolution can do almost anything with beaks. Archaeopteryx, the proto-bird, had a bony head and teeth, being a half-and-halfer. But modern birds have beaks made from keratin, the same stuff that feathers (and for that matter fingernails) are made from. This is weirdly strong and fantastically light, and it has allowed birds to come up with the most extraordinary variety of shapes and sizes and ways of making a living.

So sit with me on the banks of the Luangwa River on a bright morning and count the number of beaks: count the different ways in which a bird can catch a fish. A grey heron, the same species as found in Britain, holds still and

---

* The toucan family, the Ramphastidae, comprises 40 species in five genera.
† A joke stolen from *Finnegans Wake*.

stares at the water, waiting for a chance to lunge with a spear-shaped beak and grab. This technique works so well that there is a great radiation of herons across the world, and many of them are found in the Luangwa: the tiny green-back heron clambering in overhanging branches, the rightly named Goliath heron, the little and the great white egrets,* the black egret that adds a refinement to the wait-and-grab technique by making a shade over the water with its wings, and the black-crowned night heron that has the river mostly to itself when the daytime fishers are in bed.

Then there are kingfishers: the tiny malachite kingfisher that fishes from a perch, the chunky pied kingfisher that hovers, and so is freed from the ties of the bank, and the football-sized giant kingfisher. They have beaks of different sizes, each one adept at taking fish of different sizes. The yellow-billed stork adopts a wait-and-see tactic, standing in the water with big yellow bill agape, locating fish by touch. Behind them, immensely busy, looking like a husband showing his wife how the hoovering should really be done, is the African spoonbill, sieving up small scraps of life with its damn great ladle of a bill. The African fish eagle prefers to swoop on a fish from a riverside tree, sweeping it up in a killing clasp and taking it to a perch to butcher it with its great hooked bill. I have seen a fish eagle catch a fish too heavy to fly with: it just about got it to the bank, but couldn't take off, and eventually lost the prize to a croc. In the night, the same fishing tactic is pursued by the massive tangerine-coloured Pel's fishing owl, which generally works from the banks. Pelicans are not often seen on the river, but can be found on the lagoons when the dry season ends, where they use their preposterous bills a bit like purse-seiners. They are particularly fond of the time at the end

---

* Why are egrets white? It makes them stand out from the surrounding countryside with uncomfortable clarity. But if you look at them from the water upwards, as a fish does, then you see only the egret's pallor against the pallor of the sunlit sky. A fishing egret is close to invisible to its prey.

of the dry season when the lagoons shrink drastically and the fish are exposed. The lagoons become muddy puddles boiling with fish, and taking them out really is like spearing fish in a barrel.

And then along the river, uncommonly graceful, comes a pair of African skimmers. They look like enlarged terns – until they start to fish. Then they drop the lower mandible of the bill – noticeably longer than the upper – and plough a furrow across the water, leaving a beautiful, long, straight, ever-vanishing line, as if they were drawing across the surface of the water with a Chinese calligrapher's brush. But it's not art, not from the skimmers' point of view: their senses are fully engaged and they are ready to snap up anything they come across. And if you make a trip across Zambia to the Bangweulu Swamp, you might be lucky enough to find a shoebill: a heron with a beak like a Dutch clog that it uses to catch lungfish.

So many fish: so many ways of catching them. So many birds: so many beaks. The Luangwa riverside is a virtuoso display of evolution: an almost showing-off demonstration of how many variations can be worked on a theme. Evolution takes keratin and moulds it into beak after beak after beak, as a child makes endless shapes with a chunk of Plasticine. The feather is the soul of the bird and is common to all birds. The beak is a more individual thing, something that often defines a species. A feather is a symbol of the great class of Aves. The beak is a badge of individuality, sported by each separate species.

****!

Oh fuck, another phylum. It's probably the most famous remark made by the palaeontologist Simon Conway Morris, and no doubt the one he most regrets. All the same, I expect we can all empathise with it as we meet the phylum of Gastrotricha: the hairy-backs. Conway Morris was widely celebrated for his work on a fabulous collection of Cambrian fossils found in a rock formation called the Burgess Shale. Stephen Jay Gould, met before in these pages, made a hero of Conway Morris, and suggested that his discoveries demonstrated that Cambrian life was actually more diverse than the life of the present day. Conway Morris's[*] remark was a reflection of the astonishing hyper-diversity of the Burgess Shale fauna. However, he disagreed with Gould's conclusions, and suggested that most of the fossils could be properly accommodated in the modern phyla;[†] a view that is current orthodoxy.

---

[*] Conway Morris is both a Darwinian and a Christian, and has given lectures to make the point that the two are not incompatible. Christianity does not require its adherents to accept creationism, and science does not require scientists or anyone else to reject religion. Here he is in agreement with Gould. Gould was not religious, but wrote that science and religion were not mutually exclusive. He coined the rather orotund phrase – he was increasingly prone to such phrases in his later work – of "non-overlapping magisteria". Briefly, science can demonstrate the truth of Darwin's theories, but science cannot, by its nature, have a view on the question of whether or not a human has an immortal soul. It is not science's job to legislate on matters of belief. Such things cannot be proved or disproved and are matters of faith. Science does not deal in such things: they are not the business of science. In other words, it is impertinent for a scientist to lecture us on the non-existence of God as it is for a vicar to lecture us on the falsity of evolution.

[†] Conway Morris named one Burgess Shale creature *Hallucigenia*, so bizarre did it look. Gould believed that it was not related to any living creature; the modern consensus is that it is related to the arthropods, the great group (we shall come to it eventually) that includes insects, spiders and crustaceans.

It has to be admitted that the modern phyla don't seem to be a particularly limited bunch, and that the more you look for life – even on the huge scale of the phylum – the more you find. Or to put that another way: oh fuck, here are the gastrotrichs. Bring on the hairy-backs.

They are astonishingly numerous, that's the real point of them. They have been found at a density of 150 individuals in 10 square centimetres, say, 4 square inches, of water. Microscopic, then, found in marine and fresh waters, and most likely to found in between particles of sediment. They are also found in the film of water around grains of soil. Small things, things almost meaningless to the human mind, they are nonetheless, by virtue of their numbers, an important link in the food chain that sustains us all: feeding on impossibly minute scraps of life like microalgae and bacteria, and in turn preyed upon. There are over 700 species described, and they come in two main groups: strap-shaped hairy-backs and tenpin-shaped hairy-backs.

I have not, I suspect, thrilled my reader very deeply with this chapter. I haven't sold the gastrotrichs over-well. And yet there they are, millions and millions and untold billions of them. So many: would the world be able to carry on without them? Would the world be very different if they didn't exist? I suspect it would.

Look, no stabilisers

I have flown with eagles. I have to admit, though, that I was very frightened before take-off. It was the instruction to empty my trouser pockets that did it: to leave the Swiss Army knife and the pocket magnifier and the small coins behind, lest they work their way loose in flight and crash into the prop and break the damn thing. The prop, you see, was basically sticking out of my arse. Must be pretty fragile if you can break it with a 10p coin. That instruction rather underlined the fragility of the whole business of being up in the air. Humans aren't really meant to do it. It's against nature.

John Coppinger is an old friend; he runs Tafika Camp in (you've guessed it) the Luangwa Valley. And he has a microlight: one of those Heath-Robinson aircraft that is a strange coupling of kite and lawnmower engine. John sits in the front and holds a bar attached to the wing, which is nothing more than a triangle of fabric above his head. You ride pillion and hold onto – well, nothing much, actually. My old friend Chris Breen (the one who attended the death of the elephant), alarmed when John had gained some serious height, confessed that he was once reduced to clutching hold of the pilot. You rest your feet on things like aluminium cigar tubes.

The first time I went up, I took my place with the copper taste of nausea in my mouth. This feeling lasted all through the flight preparations and the run and bounce down the brief baked-earth runway. It vanished the instant, but the very instant, we left the ground. All at once there was the mad river beneath me, thrashing about from one side of the valley to the other like a wounded snake, as it has done for countless millennia. Down there I could see the vast mammals, now perfectly accustomed to the noise periodically made by the little aircraft: elephant, eland, impala, puku. We could visit the crowned cranes on their breeding grounds, in the soft swamps where no human can tread.

And we flew with the bateleurs. The bateleur eagle* is the great emblematic bird of the Luangwa Valley. They can be seen over most open savannah country in Africa, but they are or seem to be especially numerous in the valley. They hold their wings in a sharp V, a marked dihedral, to use the correct term from aerodynamics. This gives added stability in the roll axis. The birds are still not completely stable, and they need to make constant small corrections and counter-corrections. The word bateleur is French, and means a tumbler, a showman, a circus performer. This bateleur is a tightrope walker, tiptoeing his way across the sky with insolent ease, knowing that the air is his servant, not his master, and that gravity is in pretty much the same situation.

Flight is the birds' great advance over the rest of the vertebrates: only bats can rival them for powered flight, and they are not all that close when it comes to comparisons based on pure flight. Sonar is a great trick and bats are masters at it, but when it comes to flying, birds are on a different level.

Flight is the birds' evolutionary triumph, whether you are a ground bird that leaps effortfully into the air (like a peacock) at the approach of a predator, whether you are a small bird of the forest canopy that has forgotten that the ground ever existed, or whether you are a bateleur whose days are spent riding the air and, very occasionally,

---

* To be strictly accurate, bateleurs are not true eagles, like the golden eagle and the bald eagle. They have no trousers: no feathery covering on the upper legs. They are classified in a separate but related group called snake eagles.

flapping. A bateleur will spend eight or nine hours a day in the air, and will cover an area of around 250 square miles, 650 square kilometres. It does this with a minimal expenditure of energy. Flight is very expensive, but gliding is highly economical: you just sit there and let the wind do the work. A bateleur will soar on thermals, rising columns of heated air, and will hoard the height like a miser, relinquishing it grudgingly as it cruises across the valley in search of food: pigeons, sandgrouse, snakes.

And here's a rum thing: the bird seems to have no tail whatsoever. The red feet extend beyond the base of the truncated tail. The flight silhouette is so singular that it's a standard tease to tell first-time visitors that "this is the only bird in the world that flies backwards".* Tail-less flight is supremely efficient for all day gliding: less drag from the tail means that the bird loses very little height as it travels forward under gravity. But the technicalities of tail-less flight are hard to master, and that's why a subadult bateleur doesn't look much like a bateleur. Instead of being boldly picked out in black and white with a startling red beak and legs, this younger eagle is all shades of brown, and it has a tail. For the first seven or eight years of its life, it has a tail much in the manner that children have stabilisers on their bikes. The difference is that the tail doesn't get removed all in one brave decision: rather, it diminishes as the bird shifts through its plumage phases. The young eagle has to learn how to be a bateleur. Mastery of the airways doesn't come cheap: it's a long, hard education. Flight is a life skill that must be acquired and worked on. The adult bateleur's casual confidence comes from a ritual 10,000 hours.

Flight is what birds excel at. You can choose many species – practically any species – to spell out this mastery: the daily miracle of the bateleur will do as a starting point. There are other birds with greater speed, greater drama, greater

---

* Hummingbirds really can fly backwards, and we'll come to them in due course.

endurance, greater range and at least equal aerial skill, and we will touch on a good few of them as we continue our circling journey through the vertebrates. I remember looking down from the microlight, looking past my own left boot, at a bateleur who was flying his own flight without needing to make a horrible noise, with no need for an arse-shaving prop, out-turning us with the subtlest shift of those perfect aerofoil surfaces. The bateleur stands for the perfectly insolent nature of the birds' conquest of the air. The bateleur is the bird that knows the sky loves him.

7. BATELEUR

The crypto-bums

Entoproct means "anus inside". And that is perhaps the most fascinating things about this phylum: they are creatures with hidden bottoms. They are quite easy to overlook – the entire creature, I mean, not just their bottoms. The biggest of them measures around 7 mm, 0.28 inches. They are goblet-shaped, so they are sometimes called goblet worms, for all that they are not a lot like worms. They have also been called nodding heads.

They are found in water; all but two of the 150 species described live in the sea. They have a crown of tentacles around the rim of the goblet, and they use beating cilia to encourage particles of food in the general direction of their mouths. I know you've heard that one before: it's a very popular way of making a living. So many creatures, all of them clearly and sometimes spectacularly unrelated, have come up with the same basic solution to life's first problem of getting enough to eat: just hang about and wait to see what the sky will drop in the way of food, and help it on its way by sort of beckoning to it. It's a way of living that prompts two more or less simultaneous thoughts. The first is that a completely stationary lifestyle seems unnatural to us humans, but as we have seen, many, many creatures across many different phyla find it suits them perfectly. It's not as if hanging about all day and all night in the same place is energy-expensive: and the business of making a living is based on balancing the books. If you rampage about all day and/or all night like a hyena or a bateleur, or if you are an insanely active shrew, you need to take on a fair bit of nourishment to keep going. A shrew will famously eat his own body weight in a day. Finding food tends to be an expensive business for birds and mammals: they need a lot of food in order to be able to search for food. But if you are an entoproct, hanging about and doing the occasional bit of wafting, you can exist on much less.

The second thought is that even granted all this stuff about the economics of survival, there must still be an awful lot of floating, drifting, tumbling bits of nutrition in the water, in certain places, to support all these waiters and hopers and wafters. The larvae of the entoprocts are free-swimming, but then they settle down, in the most final sense of the term, to grown-up life, swivelling through 180 degrees and turning both mouth and anus upwards. They can propagate by cloning, but they also do sex. They tend to start off as simultaneous hermaphrodites but some species later divvy up into sexes. And that's it: wafting goblets with secret bottoms: creatures who have as fully evolved and effective a way of life as the president of the United States.

Same bat time, same bat hawk

Let's look at some of my greatest bird-of-prey moments. Why not? They include many of my best moments for all wildlife. Or all life, for that matter. Looking *down* on a lammergeier cruising along a Spanish valley, being almost knocked off my horse by a hunting sparrowhawk, the marsh harrier on my honeymoon,* the hobby that stooped on a flock of house martins by a church in Barnet,† the hawk migration in Michigan,‡ the majesty of martial eagle, the daintiness of a merlin perched on my naked fist,§ the sheer speed of a peregrine... but I could go on forever and before long I would be showing off rather than explaining the wonders of the wild world.

The drama of them. Let's have a bat hawk. Few animals on earth do drama quite as a bat hawk does. They are African birds that have cracked a specialist niche. As a result, they are busy for just a few minutes every day: but those few minutes are spectacular. At dawn, and especially at dusk, a bat hawk turns from a small black statue into a high-speed high-stealth killing machine. As the bats come out to feed at the end of the day, and as the bats return home in the fore-dawn, so the bat hawk preys. And that's

---

* Recounted at length in *My Natural History*.
† As told in *How to Be a Bad Birdwatcher*.
‡ In the American edition of the above; birds of prey encounters tend to have a tremendous significance.
§ I was visiting the falconers at Banham Zoo in Suffolk.

how you see them: as a silent clap of thunder. An expanse of still water is best, for you have a clear view and plenty of ambient light. There are also many insects associated with the water, and they bring in the insect-eating bats. It's the moment when the sight-hunting tactics of the hawk and the echo-locating techniques of the bat overlap, and it's the bat that tends to pay the price. Out of a dark sky, the darker bat hawk will strike: the shadow of a shadow: appearing in your vision for about two seconds before vanishing... leaving an indelible memory on every human tuned into such things. Remember when we saw the bat hawk? You mean the one by the lagoon at Nsolo? I'll say I do.

Birds of prey have a deep and rich meaning for humans. They are all of them glorious: and that's because they have to be. Their way of life is intensely demanding: therefore they are right on the limits of evolutionary possibility. They have to be able to stay in the air all day like a bateleur; or they have to be as swift and as secretive as a bat hawk, or they have to be as fast as a peregrine or as manoeuvrable as a sparrowhawk. They have no choice. For a bird of prey, perfection is the only option. That is, in fact, a kind of heresy: as said before in these pages, evolution is not about seeking perfection, it's about finding a solution to the questions that life asks. All the same, birds of prey look very much *like* perfection: and it's probably fair to say that they are as perfect as anything that evolution has come up with. To human eyes, they are an embodiment of everything that we find awe-inspiring in the wild world.

The notion of "birds of prey" is a fairly rough-and-ready concept. Plenty of birds kill and consume their fellow vertebrates and are not counted as birds of prey: an osprey preys on fish and is a bird of prey while a heron does the same thing and is nothing of the kind. Skuas, penguins, storks and kookaburras, to name but a few, are predatory birds that are not birds of prey. The tendency is to lump together five more or less related groups of fierce birds and call them "diurnal birds of prey". They all have superb eyesight, claws specialised for grasping and killing, and a butcher's hook for a beak. It's wiser not to speak of "nocturnal birds of prey"; these are more frequently referred to as "owls", and we'll have a look at them in the chapter after next.

These diurnal birds of prey are almost all fabulous fliers, but they have many different kinds of different kinds of fabulosity. The marsh harrier I can observe from the window of my study as I look out over a stretch of marsh is very good indeed at flying extremely slowly: quartering a feeding ground at speed that would cause many birds to stall and fall out of the air. Slowness gives it time to survey with immense precision: the better to examine what is below and when the moment is right, kill it and eat it.

The peregrine goes for the opposite strategy and is the fastest bird in the world. A peregrine, streamlined, powerfully muscled and with swept-back wings like a fighter plane, is capable of reaching 60 mph, 100 kph, in straight and level flight: only a few birds can match or perhaps beat this, some waders, ducks and pigeons. A swift, the white-throated needletail, has been clocked at 69.3 mph, 111.6 kph. But in a stoop – a thunderbolt attack on a flying bird, plunging from a height and so exploiting gravity to add to its speed – the peregrine doesn't so much speed up as pass into another reality, reaching 200 mph, 322 kph. Even at such speed the bird has enough control to home in on a flying bird, one that is manoeuvring in three dimensions and often enough, attempting to escape – though plenty are caught completely unawares.

The techniques of all-day gliding and soaring* are practised by vultures of both the old and the new world.

---

* Gliding is unpowered flight and therefore, by definition, downwards: a good glider is the one that falls least quickly. Soaring is the unpowered gaining of height. You can watch seagulls using the updrafts from cliffs, and many birds (not just birds of prey) will spiral upwards on a column of warm and therefore rising air, which is called a thermal.

These two groups are not, in fact, closely related but they are both adept scavengers. The new-world birds, which include the famous condors, have, unusually in birds, a sense of smell. The sparrowhawk is a hider and ambusher, the greatest of all sneak-strikers, with almost impossible manoeuvrability. I have seen a sparrowhawk fly at full speed though a wood, going through gaps a blue tit would think twice about, folding its wings and unfolding them on the far side of the gap to emerge and strike at unsuspecting birds on the other side, often inverting in the air to stretch out a single foot armed with killing claws. The osprey is in a group by itself: the dashing fisher that is found all over the world.

And one oddity. The secretary bird is a bird of prey that hunts on foot, stalking about the open plains of Africa apparently with its hands behind its back, in the same terribly interested fashion of Prince Philip reviewing the troops. It specialises in taking snakes from the ground.

Birds of prey have a special meaning for humans: often a bellicose and hyper-masculine one. Best not to explain, perhaps, that among birds of prey, the female is almost always larger and stronger than the male and capable of dealing with bigger prey. Birds of prey are the birds men would like to be: in America's National Football League, three of the 32 teams are named for birds of prey: Philadelphia Eagles, Atlanta Falcons and Seattle Seahawks. Birds of prey are the most admired birds on the planet: and they are also the most persecuted. The illegal killing of birds of prey is one of the most intractable problems of conservation. In England, only one pair of hen harriers succeeded in breeding in 2012, despite the fact that there is capacity for more than 300; the following year not a single pair bred. The problem is that most hen harrier habitat is on grouse moors that are managed for shooting. A satellite-tagged bird, Bowland Betty, was shot a few weeks after she was tagged. When it comes to perfectibility, we humans time and again succeed in bringing off the perfect contradiction, particularly when it comes to our relationship with the wild world, which we love and hate in equal measure.

Lobsterisimus bumakissimus

Oh dear. Back to bums again. That's nature for you: never far away from the next bum.

"The worst job I ever had was with Jayne Mansfield. I had the terrible job of retrieving lobsters from her bum." This from the notorious album *Derek & Clive Live*, by Peter Cook and Dudley Moore, released in 1976. It was 19 years later when what is arguably an even worse job involving lobsters was discovered by the scientists Reinhardt Kristensen and Peter Funch. It was also the discovery of a new genus: a genus that didn't fit anywhere else in the Animal Kingdom and so constitutes its own phylum. These creatures were discovered on the mouthparts of the Norway lobster. Related creatures were subsequently found on the mouth-parts of the European and American lobsters. It was the first phylum discovered since 1983, when the loriciferans[*] were first described; this phylum is called Cycliophora.

They are also called symbions, for their interdependent life-style, and they go through two main life stages. They begin with an asexual feeding stage (like a caterpillar) when they take small shares of the meals consumed by their pet lobster: the lobster works for them as surely as the lobster of the French romantic poet Gérard de Nerval did for him. (Nerval used to take his pet lobster Thibault out for walks[†]

---

[*] See page 200
[†] "Why should a lobster be more ridiculous than a dog? Or a cat, or a gazelle, or a lion, or any animal that one chooses to take for a walk? I have a liking for lobsters. They are peaceful, serious creatures. They know the secrets of the sea, they don't bark, and they don't gnaw on one's monadic privacy like dogs do."

attached to a blue silk ribbon in the Palais Royal.)

The symbions don't do much walking, though; they remain attached to the mouth by a sucker, taking advantage of their host, another sucker. They can propagate by budding. But they will also move into a second life stage, and become either male or (you've probably guessed it) female. The males abandon such things as mouths and anuses (bums, if you prefer) and set off as free-swimmers to seek a budding female, which they will then impregnate. The female subsequently separates but she will retain her digestive system, although digestion is no longer its prime function. Impregnated, she leaves her host and finds another. As she does so, her digestive system morphs into a larva, which escapes from her when she dies. The sex cycle is triggered when the lobster moults its skin in order to grow.

Just a few of them, then, in just the one genus, so far as anyone has discovered. There may be many more, but that requires an awful lot of looking at very small things, and not many people are qualified to do it. For most of us, it would be the worst job we ever had. All the same, people like Kristensen and Funch are in the same position as the space pioneers of fiction, travelling beyond the human experience to find new creatures living in a manner previously unimaginable. How alien – that is to say, remote from human ideas of life – can you get in the Animal Kingdom? Here is a creature as different from an earthworm as it is from a human being, living its strange but practical and effective life in the mouth of lobster, the female with a gut that turns into her own child, but only after she has been impregnated by a bumless male.

The dark side

Owls are not all that closely related to the diurnal birds of prey we've been talking about. Some say they are closer to nightjars, which are nocturnal fliers after insects. They have come down a separate evolutionary road, and their ripping beak and killing claws represent another convergence. Primarily, they have evolved for darkness. So much of evolution is about seeing a gap in the market and filling it, and for an owl, the USP* is that they can work in the dark. Not the pitch dark of a cave, as a bat can: but when it comes to working in low light, they are the best birds in the world. The diurnal birds of prey have the hours of daylight to themselves – and a bat hawk and some others can exploit conditions at dawn and dusk – but the true night is for owls.

There are 249 species in the superb *Owls of the World*,† starting with barn owl and ending with marsh owl. I've seen both, though alas, there are a fair few in the middle I haven't; Usumbara eagle owl and Maria Koepcke's screech owl, to name but two.

Owls all have the bright two-eyed face that reminds us of ourselves. Perhaps that explains why we equate them with exceptional intelligence.‡ Owls are about seeing in poor light, they are about hearing, and they are about silence. That is the triple package that defines them and the way

---

* Unique Selling Point.
† *Owls of the World: A Photographic Guide* by Heimo Mikkola.
‡ Though Hindus traditionally see owls as stupid. And Owl, in *Winnie-the-Pooh*, pretends to be clever while actually being rather daft. He signs himself WOL.

they make a living. The forward-facing eyes give them binocular vision, and therefore an acute three-dimensional image of what is before them. This operates over a relatively narrow field of view; this stereoscopic view covers less than 50 degrees. The eyes are fixed in their sockets, so to compensate, owls have a disturbing, and really rather human way of turning their heads.

This becomes superhuman relatively quickly: a long-eared owl can turn its head though 270 degrees; it has 14 neck vertebrae to pull off this trick. The African scops owl, among others, has false eyes on the back of its head, so its potential prey can't tell which way it is looking. The eyes are widely spaced. Owls improve their three-dimensional vision by bobbing their heads comically to get a fix on what they are looking at.

An owl's hearing is reckoned to be ten times better than a human's. It is helped by the wide facial disc, which focuses the sound, in the same way that we can cup our hands round our ears to hear better. Most birds hear through two small holes in the skull; owls have a huge half-moon shaped vertical slit. Some species have their ears set asymmetrically, which allows them to get a cross-bearing on the source of a sound: snowy owls can drop precisely onto a lemming hidden in a tunnel beneath the snow. Obviously, precise hearing wouldn't be very effective if an owl's ears were filled with the sound of its own progresses, but owls can fly without making a sound. Their wings are fitted with feathered silencers. This is not only helpful for stealth – it helps that a short-tailed field vole can't hear a hunting barn owl – but also essential if the barn owl is to use its sense of hearing to get an exact cross-bearing on the creature it is hunting. Owls tend to fly very slowly, the better to examine the world through their two principal enhanced senses, and in near-complete silence. Barn owls are less committed to night than some species, and are particularly fond of dawn and dusk – crepuscular is the pleasing word to describe this, but I have often seen them hunting during the day, especially in winter, when they like to patrol the shaggy edges of a field following the line of a hedge. The field voles they look for are as hidden in the daytime as they are at night, for they move in tunnels between the stalks of rank grass. But an owl can hear them.[*]

If you are a creature of the night, no matter how good your eyes are, you're still going to find it hard to relate to your own kind, simply because you don't see them all that often. As a result, owls tend to be very vocal. They call to each other often: keep-offs and come-hithers and calls that mean I'm-here-where-are-you? This is technically known as a contact call. The land behind my house is full of tawny owls. It's autumn as I write this chapter, a season when owls are on the move: young owls must leave the place in which they were brought up and seek their fortunes. Especially, they need to seek a place where they can hunt. So at this time of year there are frenzied debates about ownership, precedence and gender. The wavering hoot from every horror film you've ever seen[†] is answered, sometimes by birds who can't really hoot properly, being too young to have got the knack. The mature and assured hoot of a full adult must be a dismaying sound to them. Tawny owls will also make a sharper two-syllable call, often transcribed as "kee-vit", with the stress all on the last syllable. This sometimes leads to frenzied nocturnal duets in which one owl's sharp contact call is answered by a territorial hoot. A tawny owl doesn't go tu-whit tu-whoo, it goes tu-whit *and* it goes tu-whoo. But there is nothing mechanical about it:

---

[*] A kestrel also hunts short-tailed field voles without seeing them very much, but it still relies entirely on its eyes. A kestrel can pick out a vole by sighting the fresh scent trails that the vole makes with its urine. The kestrel can see them because its vision takes in ultraviolet light.

[†] Most night creatures have a sinister reputation, often associations with death and sorcery. But it's not owls that humans fear, it's the dark, which was the time of maximum danger for our savannah-walking ancestors. An owl cry tells us that this is the time to be afraid.

the two calls can be uttered at differing intensity, frequency and tone, and at times, the two calls can merge.

Owls calls are always evocative. The wood owl in Africa asks: "Now then, whooooo's a naughty boy?"; the European scops owl makes a sound like a submarine's sonar; the African eagle owl goes in for a basso rhythmic grunting while a juvenile Pel's fishing owl is said to make a sound "like a lost soul falling into the bottomless pit."

Owls tell us about the versatility of the wild world. It is a mistake to think that every single possible ecological niche has a creature filling it, but nature is always prepared to have a damn good try. It came up with all those birds of prey to dominate the day, and did the same for night, and it did so in its usual bewildering number of forms, with the elf owl that weighs no more than an ounce, 31 grams, and is no more than 5 inches long, to the Eurasian eagle owl that weighs 4.5 kilos or 10 pounds.

Is-ness

What's the point if it all? In the bad days and worse nights when Winston Churchill's black dog sneaks through the window and lies fawning at your feet, the question seems unanswerable. My mother said it wasn't a dog: "I like dogs, I wouldn't mind if it was a dog. It's a black cloud." Depression is a terrible thing, and most of us know the taste of it, while some must live on the stuff for day after day: black nourishment for a lost soul trying to be found. And the worst thing that the dog or cloud can inflict on you is that sense of purposelessness. There really is no point; I'm just existing. Nothing I do matters, nothing I do will ever mean anything, nothing I do will last. I just get up and go through the motions of the day and then lie down at night and sleep, or try to. For a human it's the most terrible thought in the world. A human needs to be worth something. A human needs to matter. A human can't just live. Or if that's what we do, we need some illusion of mattering.

But as we look at the Animal Kingdom spreading all around us in the broadest way immarginable,* we are forced to realise that just being is what many of us do. A human being can sometimes find it hard to see the point of his own existence: what, then, to make of the life of a gnathostomulid? All you can say of such a creature is that it exists in order to exist. These are worms that live between grains of sand, and they are sometimes called jaw worms. Their great talent is that they can live in places with very little

---

* Another line from *Finnegans Wake*.

oxygen: in sand and mud beneath shallow coast waters, where they can operate their second great gimmick, their jaws and teeth. They are the Father William worms – he was the old man in *Alice*, who ate up the goose "with the bones and the beak". He explains that in youth, he took to the law

> "And argued each case with my wife,
> And the muscular strength that it gave to my jaw
> Has lasted the rest of my life."

The muscular strength of the jaws of the gnathostumulid allows them to gnaw living matter from grains of sand. They are simultaneous hermaphrodites, perhaps a consolation in what seems to be rather a restricted life.

But all lives are restricted: or all lives are meaningful. Gnathostomulids live and reproduce and die: and that's glory enough. Perhaps a creationist can tell us why God made them, a dull creature for a dull day. But what is relevant, to them at least, is that they live. There are more than 100 species of them, and there are probably very many more, if we could be bothered to look for them, but gnathostomulid experts are thin on the ground compared to birdwatchers.

Life comes up with these strange, and to us, pretty pointless creatures, not to show off or even because it can, but because it does. The forces of life make creatures and what these creatures do is life. If they become ancestors, then the line continues. Students of Zen poetry talk about the ah-ness of things. This is a concept better shown by example than explanation. Here is a classic example of ah-ness from the Japanese haiku master Basho, a poem about flowers.

> The ominaeshi, ah!
> The stems as they are,
> The flowers as they are.

It reads a trifle awkwardly, I know, but no doubt it flows in Japanese. The idea of ah-ness is about explaining in words that the matter in question goes too deep for words: so it's a canny little paradox, the grace of words conveying the emotion of wordlessness. A haiku poet often experiences such a moment when a cuckoo announces the arrival of spring.* I have touched on the same sort of emotion in recent chapters when I have written, or tried to write about, encounters with a bateleur and a lammergeier.

So let's move from ah-ness to is-ness, which I think is a related phenomenon. Is-ness is shared by all living things. A gnathostomulid is because it is. It requires nothing more than existence to be worth existing. Life is about being alive: life is what life is and because life is. We can say with the gnathostomulid: "I is therefore I am."

---

* The cuckoo in question is the lesser cuckoo, *Cuculus poliocephalus*. We shall get to cuckoos in due course as we circle through the birds. The lesser cuckoo is called in Japanese *hototogisu*, a fine word that few haiku writers have been able to resist. Here's an 18th-century example from Kagami Shuko, a fine poem about the frustrations of birdwatching, trying and failing to get a good diagnostic and above all tickable view of a cuckoo and failing.
     Hototogisu – can't get my head through the lattice
Both these poems are taken from *The Classic Tradition of Haiku*, edited by Faubion Bowers.

Crisis relocation

Crisis relocation is a bit of jargon from the Cold War and it means being somewhere else when the crisis happens. If there are a hundred nuclear missiles heading for Washington it's a good idea to get the powerful people in government and administration away: preferably a good while before the missiles are on their way down. It's a principle of anticipation. And though the terminology dates from the 1970s, the concept goes back to the dawn of life. Whenever a crisis strikes, the best place to be is far away. Just as a wise batsman plays the demon fast bowler from the non-striker's end, so the greatest strategy for avoiding a crisis is not being there when it happens.

The better you can fly, the more adept you are at crisis relocation. That puts birds ahead of any other class of animals on the planet. Winter is a crisis that strikes the planet once a year in every place outside the tropics, but not in all places at the same time. There are around 10,000 species of birds: something like 1,800 specialise in crisis relocation. That is to say, they are long-distance migrants. When winter strikes and there is nothing to eat and the killing chill settles over the once-friendly and fecund expanse of countryside, they are somewhere else. And of the remaining 8,200-odd, many, if not most, take smaller journeys, often gathering in peripatetic flocks that follow the food sources, flocks that break up once the warm weather arrives again and it is time to think about breeding and to stop being one of many and start being half of a pair.

So let's talk about racehorses. It's hard for me to explain, bit by bit, detail by detail, the difference between a superfit racehorse and a nice horse that likes a bit of a gallop. I could talk about relative length of leg, about the different build, about the fineness of the head, above all, about the way it moves, the way an ultrafit horse will overtrack – that is to say, it will plant its rear hooves some inches *ahead* of its forward hoof-prints. But that's not what I actually look at when I see a top horse. I see a sort of elasticity: a stretchiness of movement that indicates the trained-to-the-minute athleticism of a thoroughbred racehorse. It's not so much an analysable quality as a vibe. That's not a horse, *that's* a horse.

So what is the difference between a common tern and an Arctic tern? Both are lovely and graceful birds. The Arctic has shorter legs, longer tail streamers, no black tip to its bill, translucent wings. Not that I've ever really noticed. When you see an Arctic tern and you say, *that's* a tern, then you've got the idea. Elasticity, yes, especially in the way the body rises and falls with each wingbeat, as if to get maximum value from each stroke, the sense of a flight more delicate, with faster and shallower wingbeats and shorter and narrower wings.

The common tern is a mere long-distance migrant. The Arctic tern might as well be called the Antarctic tern, or perhaps the bipolar tern. Arctic terns don't just drop down to Africa for the winter: they travel all the way down for the rich pickings of the Antarctic summer, and then come back again. They nest in the north, all around the pole. North Scotland is included here, and they have bred as far south as Brittany and Massachusetts. When breeding is done they set off on the travelling life again, along coastal migration routes that follow the shape of the continents; some will cross the Atlantic at the shortest possible (how do they know?) bit. They don't make straight-as-the-crow flies journeys: bugger crows; these are Arctic terns and they take rambling meandering routes that pick up all kinds of food sources and even food bonanzas on the way. It's been estimated that they cover 70,900 km or 44,300 miles in a single year.

Mad for daylight, these terns must be. I'm from southern England, so I'm always staggered by the length of the day during the summer in northern Scotland. If you go there to look for wildlife you can spend a decent period of time without seeing darkness at all: night is really a brief twilight they call the Summer Dim. This small and marvellously athletic bird spends most of its life in the daylight.

Distance means less to a bird than it does to any other creature. Some mammals, such as whales and wildebeest, migrate, and so do some insects, like monarchs and painted ladies. Some species of bats do it as well, but most species prefer to hibernate, a more economical strategy for dodging the winter. Some even do both. But birds are the best when it comes to the great survival art of being somewhere else.*

Certainly, a migration is full of dangers, but then so is staying in the same place. The pole-to-pole strategy is the ultimate form of the game. Hardly anything can survive in the polar winters, but if all birds were residents, the annual bonanzas produced by the warm weather would go untapped. Migration is not just about avoiding harm: it's also about taking advantage of things that are only available to the highly mobile.

A tawny owl, as discussed a couple of chapters ago, will spend most of its adult life in the same chunk of woodland, learning every inch of it, knowing every hunting perch and everything that can be caught from it. As you can find your way across your own house in the dark when you take a midnight pee, so a tawny knows its wood, processing meagre scraps of shadowy information in a meaningful way. It will

---

* The common poorwill, a nightjar relative that lives in the United States, does actually hibernate. A number of species – some swifts, nightjars and hummingbirds – go in for torpor, a short-term drop-out when conditions are against them. These include the swifts that come to Britain for the summer.

use its wings for silent descent onto unsuspecting prey. But other birds have their being in a much wider world and use their powers of flight to conquer distance. It's only in the last generation that humans have been able to think of the world in the same way.

The Quaker worm

The Gospel according to Matthew begins with a genealogy: "Abraham begat Isaac; and Isaac begat Jacob; and Jacob begat Judah and his brethren; and Judah begat Pharez and Zerah of Tamar; and Pharez begat Hezron and Hezron begat Ran…" We are all tied to the past by the genes we carry in every cell of our bodies – and we humans have as our ancestor the same creature that begat xenoturbellids.

Xenoturbellid means strange flatworm: same prefix as xenophobia, which means fear of strangers. Here's a small and apparently pointless little phylum of apparently pointless little creatures. Oh – and by the way, they're related to you. Not just in the sense that everything in the Animal Kingdom has some kind of relationship to everything else in the Animal Kingdom. These weird little lumps of nothing very much are really quite close to us on the evolutionary bush. The branches keep on forking: but the common ancestor shared by humans – by all vertebrates – and the xenoturbellids was on the same stem for far longer than you'd believe possible to look at them. The fork where we parted company is not as far behind us as we'd think.

If you ever ask a Quaker what Quakers do and what Quakers believe, you will generally get told all the things they don't do and don't believe in. Xenoturbellids are in much the same position. They have no brain, no through gut, no excretory system and no sex organs in any coherent form. They really are not anything much, but they are part of our comparatively recent history just the same. What have they got, then? They are wormy things about 4 cm,

254

1.6 inches, long and they have flagellate cells – cells with little whips on, a reasonably familiar concept by now. They can be found in waters off Scotland, Iceland and Sweden. They were discovered in 1915 by Sixten Bock, a name I shall surely use for the villain if I ever write a thriller. They were not fully described until 1949, when Eimar Westfald did the job. There are two species described so far in the entire phylum: they are called *Xenoturbella bocki* and *X. westfaldi*, as is only right and proper.

Their relatively close relationship with ourselves was put forward by Max Telford of Cambridge University, who said: "We have been able to show that amongst all the invertebrates that exist, Xenoturbellida is one of our closest relatives. It's fascinating to think that whatever long-dead animals this simple worm evolved from, so did we." Relatedness happens. It's the way life works.

Think what that means for a moment. The ancestor shared by xenoturbellids and humans was presumably no more complicated than the Quaker worm itself. In other words, it was an insignificant blob of nothing much. Perhaps, like the Quaker worm, it had no brain and no bottom. The ecology of planet earth would hardly have been changed forever if the xenoturbellids had gone extinct and left no further ancestors. So: what if the same fate had overtaken the xenoturbellids' ancestor? The one they share with us? Say that a worm's ancestor didn't actually become an ancestor: well, that would hardly constitute a great event in this history of life on earth, not like the KT extinction that finished off the dinosaurs. If some great scientific intelligence had been tracing the planet's history, this small extinction would have been a footnote[*] at very best.

And yet without that common ancestor, fish wouldn't swim, bats wouldn't fly and dinosaurs would never have

---

[*] Not the most important part of a book, no matter how much the author may love them.

ruled the earth, Arctic terns wouldn't be migrating from pole to pole, humans would never have walked the savannah, Joyce would not have written *Ulysses*, Darwin would not have written the *Origin*, I wouldn't have written this book and you wouldn't be reading it. We're here because of a worm's ancestor.

Swift scramming frenzy

Birds are about flight, and that makes swifts the ulti-mate birds. Nothing flies like a swift. There are species of swifts (and swiftlets and needletails) in every continent bar Antarctica, but let's concentrate on the one best known in Europe, the one we call simply swift, or European swift when we need to stress the differences.

> Their lunatic limber scramming frenzy
> And their whirling blades
> Sparkle out into blue –

Words of Ted Hughes. The pub opposite a former house of mine has a seat in the car park that was used once every year. That was when I and my older son sat there in late July to watch the swifts screaming up and down the street. They filled the air with low-level hooligan runs, racing each other for the sheer fun of it, like kids on motorbikes, for these were non-breeders with a belly full of food and nothing to do but raise a little hell. How long do you think they spend in flight?

It's an answer that can compel the most bird-indifferent person in the world into brief wonder. Most swifts won't perch* at all between fledging and breeding – and that can be four years. A flight four years long: that puts the economy-class flight to Sydney into perspective. I've said that distance

---

* And anyway it's not really perching as a sparrow understands the term – just clinging onto a vertical surface or going bellydown on a horizontal one.

doesn't mean much to a bird: it means almost nothing at all to a swift. It's been estimated that swifts cover 500 miles, 800 km, in a day. If the weather is bad in England, they can always drop over to, say, the Bay of Biscay, to load up on the aerial plankton that fuels their endless flights. They roost on the wing, circling for hours at high attitude in the thin, gelid air.

Their biennial migration to Africa is almost the least of their journeys. For them flight is not one big heroic effort: it's a constant way of being: *volo ergo sum*. Their lifetime distance has been estimated at 1.28 million miles, more than 2 million km. And in straight-and-level flight, they are fast: young swifts flying competitively have been clocked at close to 70 mph, more than 110 kph. I have tracked swifts at 60 mph, 96 kph, from a vehicle. They are so deeply committed to the air that their scientific name, *Apus apus*, means no-foot no-foot: it was believed until comparatively recently that they didn't haven't any feet at all. Their feet are pretty tiny: just a few claws. They often nest in roof spaces (holes in trees being hard to find in cities) and I have seen them belly-shuffling across a roof to creep in under the pantiles. It is an abiding myth that they can't take off from a flat surface: it's possible but tricky. Adults can do it, but young birds, particularly those just fledged, will sometimes belly-out and strand themselves on the ground, and appreciate a helping hand to toss them back into the air. Taking off isn't the most important skill for a swift: once they're up, they tend to stay up.

As said earlier, they're not related to swallows and martins; though their swirling flight and swept-back wings are not dissimilar. Both feed on aerial plankton. But the swift's wings are much stiffer and they flicker rather than beat: a very distinctive difference once you've picked it up. Their wings seem to fluoresce at the leading edge in bright weather: as if their vibrant passage were too much for the air to deal with. They seem to be all black, with the under-chin pale patch occasionally visible. And that shape, that scythe shape, there's nothing like that: "As if the bow had flown off with the arrow," as Edward Thomas wrote.

They have rather replaced the swallow and the cuckoo as the bringers of spring in Europe, says Mark Cocker in his magisterial *Birds Britannica*. Cuckoos have declined drastically, and swallows are seldom seen over cities. Swifts are more aerial even than swallows, and fly higher, and up there they find the aerial plankton that's out of the swallows' easy reach. "They're back – which means the globe's still working," as Ted Hughes wrote in the poem quoted earlier.

I have also seen their arrival in Africa. They like to surf in on the weather fronts: I remember hearing the quintessential sound of England in July, the screaming of swifts, out in the Luangwa Valley at the end of October. The ferocious flashbulb light faded as if someone had wound down the dimmer-switch; in minutes the sky turned the colour of ink and rumbled interrogatively, and in came the swifts, maybe a thousand feet up, looking as if they were towing the rains behind them. They bring the sun to England, they bring the rains to parched Africa, and they do so without bothering to touch the ground. Flight isn't what they do: it's what they are. And perhaps that makes them the bird of birds: never mind *Apus apus*: they are *avis avium*.

Gutless, brainless

Acoelomorpha means no body cavity. If you wanted to name this wormy thing after something it hasn't got, you are rather spoiled for choice. This is the super-Quaker worm. It also has no gut, no circulatory system, no respiratory system, no excretory system, no brain, no gonads and no duct associated with female reproduction. It doesn't even have any ganglia, any nerve centres. It does have a network of nerves beneath the outer covering. It also has a statocyst, which assists balance and movement. It has ocelli, which means it can tell light from dark. Some species are free-swimming, moving about with plankton; while others crawl on algae or exist between sand grains. They are all under 2 mm, 0.079 inches in length.

They are perhaps the ultimate meaningless piece of animal life, so far as a human is concerned. We can't even decide how to categorise them: it's only since 2004 that they have been considered a separate phylum, though that status is disputed. They are a classic demonstration of the fact that the world was not set up for human convenience or to aid human understanding. Acoelomorphs are there to baffle humans with their remoteness, with the impossibility of getting any kind of intuitive grasp of the nature of their existence.

We're going to get on to some super-sexy inverts very soon now, but as every married person knows, life is not just about the sexy bits. It's about life in its entirety: including the bits you don't think about much and those things you really can't be bothered with and those things whose existence you're not even aware of. But out there, far below the threshold of human awareness, the cavity-less worms are living their brainless and utterly meaningful lives, and they've never given a thought to anything as small as humanity in the brief gutless breathless expanse of their lives.

Jewels that breathe

Pineapple rock, lemon platt, butterscotch. So begins the eighth chapter of *Ulysses*. This chapter is much concerned with sweetness and sweet things, so I'm going to follow the same model. Here we go:

Green hermit, green violetear, Tyrian metaltail, greenish puffleg, bronzey Inca, booted racket-tail, rufous-tailed hummingbird, steely-vented hummingbird. That's my hummingbird list from three days in the Andes, not far from Medellín, the birds identified not by me but by the miraculous Diego Calderon-Franco.

We barely scratched the surface. There are more than 180 species of hummingbird in Colombia, and the sweet menu is unrelentingly gorgeous: fiery topaz, blue-fronted lancebill, purple-crowned fairy, amethyst-throated sun-angel, spangled coquette, black-tailed trainbearer... so many of them named for precious jewels, they catch the light as aerial brilliants, so much lovelier than the ones we cut and polish and trade and wear. I'm reminded once again of the Underlanders in *The Silver Chair*: "'Yes,' said Golg. 'I have heard of those little scratches in the crust that you Topdwellers call mines. But that's where you get dead gold, dead silver, dead gems. Down in Bism we have them alive and growing. There I'll pick you bunches of rubies that you can eat and squeeze you a cup full of diamond-juice. You won't care much about fingering the cold, dead treasures of your shallow mines after you have tasted the live ones of Bism.'" Hummingbirds are jewels from the land of Bism brought to the surface to delight us Topdwellers: jewels that live, jewels that breathe, jewels that hang in the air as if air were something you could perch on.

I've never got the hang of hummingbirds as members of a species – and there are around 350 of them altogether. Perhaps I need to spend more time birding in the tropical bits of the Americas. The trouble is that every time I see one, I'm bowled over at a far higher taxonomic level. The family Trochilidae has me so mesmerised I find it almost impossible to start looking for the diagnostic features that will allow me to separate hummingbirds into a sicklebills and thornbills and sapphires. Diego did all that stuff for me when we did our mountain birding.

When it comes to hummingbirds, I'm like a non-birder listening to a spring morning's birdsong: it doesn't matter what kind of birds they are, does it? They're all wonderful. And I, who firmly believe that the key to still greater wonders lies in looking harder and learning more, find myself hypnotised by the whiz and buzz, the flash and vanish of hummingbirds. This is doubly the case at sundown, when I am, say, at Regua* in Brazil, a rainforest project I visited in my capacity as council member of the World Land Trust. The extravagantly plumaged swallow-tailed hummingbird used to visit us every night: as I took my first drink of the day he took his last, visiting the artificial nectar feeders while I had a beer. Hummingbirds are the perfect accompaniment to beer: they come to you, flaunt themselves, and leave: you don't have to be quiet or discreet; you don't even need binoculars.

Hummingbirds have invented an entirely new way of being a bird. There's nothing quite like them. First, in terms of manoeuvrability, in terms of agility, they are the supreme aeronauts among the supremely aeronautical birds. They can hover so immaculately that their bodies look as if they

---

* Regua is the Reserva Ecologica de Guapiassu, and it's about 90 minutes' drive from Rio. It is involved in rainforest and wetland restoration and gets some financial support from World Land Trust.

have been pinned to the air like butterflies onto a board – except that they have no wings, for they are invisible apart from a blurring of the air. They can fly backwards as easily as they can fly forwards: unlike with the bateleur, this is no illusion but the real thing. They look more like insects than birds, and some aren't much bigger than a large insect: so much so that you almost feel you could swat them without remorse. The bee hummingbird is only 5 cm, a couple of inches, long.

They have chosen to live in an insect-like manner, using nectar as the energy source for a lifestyle that requires constant refuelling. And at night, they shut down: an energy-saving technique that's not far off hibernation. Their metabolism slows right down, their body temperature drops, and they lapse into a state of torpor until the sun warms them up and the process can start again.

As we have seen already, flight is extremely demanding of all birds. But hummingbirds have taken the concept of flight to extremes. They have the most energy-expensive lifestyle of all the birds on the planet. In the entire Animal Kingdom, only a few insects have a higher metabolic rate. Hummingbirds are only ever a few hours away from starvation: life is lived as a perpetual crisis. Nectar, then, is the centre of their lives: but nectar on its own is not enough. They must supplement this with protein from soft-bodied insects and spiders. They move and forage in short, sharp bursts: only about 15 per cent of any one day is spent in flight; any more would be impossible. The rest they spend perched, resting up, digesting.

Hummingbirds have co-evolved with flowers, the two parties swapping food for pollen delivery. Different species have different bills suited to different sorts of flowers, from thornbills that stab flowers at their base to drain out the nectar, to swordbills with a beak as long as their entire body for inserting into great floral trumpets: a range between 1.3 and 10 cm, half an inch to 4 inches.

Some more numbers: some species can fly at 54 kph, 34 mph; they can flap their wings at a rate of between 12 and 80 times a second; their heart rate can reach 1,260 beats a minute, and can be 250 at rest; they have a body temperature of around 41 degrees Celcius that drops to 21 at night; they can be found at 17,100 feet, 5,200 m in the Andes; their brains are more than 4 per cent of their body weight; they feed half a dozen times every hour and visit 1,000 flowers in a day. Ruby-throated hummingbirds make a biennial crossing of the Gulf of Mexico, covering 500 miles, 800 km, in one go. Before they set off they feed up and double their body weight. They need to be pretty smart (hence the brain weight) in order to establish and carry an ever-updating flower-map in their heads. Most of the colour you see on a hummingbird is not pigment: it's an effect of refracted light, like a prism or a soap-bubble. That's why hummingbirds can seem to explode into being in front of you as they catch the sun in just the right way: it also explains why so dramatic a creature can simply vanish.

I hope I have done hummingbirds some kind of justice here. After all, the *Loa loa* worm got a decent show earlier in this book, so if I am to abide by the doctrine of equal time, hummingbirds need fair representation. In the film *Manhattan*, Woody Allen's character says to Tracey, who is played by Mariel Hemingway: "You're God's answer to Job. You would have ended all argument between them. He'd have said 'I do a lot of terrible things but I can also make one of these'. And Job would've said: 'OK, you win.'"

Just one more thing…

This is the Columbo phylum. We're very, very nearly into a super-sexy new phylum of invertebrates, and with them perhaps the sexiest single species of invert that exists on the planet, in a chapter that can't help but involve both Captain Nemo and James Bond (and I'm really not building this up too much) but before we get there, well, just one more thing. Just one more phylum. Just one more obscure and grubby and forgettable phylum, just one more phylum that shows the breadth and wonder and mind-shattering diversity of life on earth.

Columbo, lest we forget, was an American detective series: Columbo himself, the man in the grubby raincoat despised by everybody, always asking the most unpromising people "just one more thing". And as a result, he invariably solved the case and brought peace and justice to the planet. So here we are, then, just one more phylum… and they're the phoronids. Filter-feeders, like so many others: all those gatherers of scraps, all those refuse-collectors, all those snappers-up of unconsidered trifles. The phoronids wouldn't thank you for confusing them with the breathtakingly uncomplicated creatures we have been looking at in recent chapters. Quite apart from anything else, they have a lophophore. This is a crown of tentacles, something that makes them something of the crown of creation in the world of filter-feeders. They live in all oceans except the Antarctic, between the intertidal zones and down to depths of 400 m. They are sometimes found in extraordinary density, tens of thousands of them to a square metre.

They live inside upright tubes made of chitin,[*] which protect their soft bodies: they can move within the tubes but never leave them. They're mostly around 2 cm long, less than an inch, though some species can reach 50 cm, not far off two feet. There is a flask-like swelling at the base which anchors the creature and allows it to retract inside the tube. Cilia on the tentacles draw food towards the mouth – and yes, these creatures are so advanced that they possess a bum; essential information, as I'm sure you'll agree. Their blood contains haemoglobin, which is very unusual in a small animal. It means they can live very effectively in oxygen-poor environments. Their larvae know what it is to be free-swimmers and express that freedom for just 20 days before they settle down and change – in the space of 30 minutes – into adults. Had I been a phoronid, I could have obliged all those teachers who ordered me to grow up, Barnes, by doing so in less time than it takes to conduct a Latin lesson. In that lightning maturation, the larva's tentacles are replaced by the lophophore and the anus (bums again, sorry) shifts from the base to a point near the tentacles, so the bottom becomes something more like the top.

So before I rush you on to the super-sexy beasts I've been talking up so desperately, let's pause for a moment and consider this great swathe of phyla we have been visiting: filter-feeders, wormy things, many of them not troubling themselves with a bum, still less a brain, not only living full and effective lives, but doing so in ways that are completely different from any other form of life on earth. Phoronids are not acoelomorphs, *perish la pensée*, and they're not xenoturbellids. It is not strictly accurate to say that the inventiveness of the Animal Kingdom is limitless and infinite: this book is trying to touch some of those limits. But the creativity and practicality of the kingdom Animalia is way beyond the limits of the human mind.

---

[*] The same substance that forms the outside covering (exoskeleton) of insects.

Have I reached the limit of these forgettable phyla, though? I'm not sure I have. Taxonomists are always rethinking the way we look at life on earth, and if they're not doing that, they're discovering new things: sometimes a new species, sometimes a new phylum. Life is in a constant state of flux and movement, and so is the way we understand it. Next week the phoronids may be gathered up into some other phylum and lose their distinct status: or they may be split into two completely distinct phyla. The wonder of looking at all these minor phyla, all these unconsidered ways of making a living, these lives so remote not just from our own lives but from our capacity to understand them, is that we get some kind of grasp of the community we belong to. Phoronids are forever strangers: forever brothers and sisters to us. We are unlike and we are just the same.

The wardrobe bird

I once thought about doing a book of wildlife epiphanies, deeply deep, naturally, though perhaps not on green oval leaves. I even got as far as taking notes: a nice notebook can have that sort of effect if you're not careful. I wrote down a few notions and recorded a few epiphanies: a shard of garish blue plastic that some idiot had dumped in a beauty spot on the South Downs – which turned out to be an impossibly blue Adonis blue butterfly. There are notes about a tiny flash of white viewed on horseback, and the pleasure of knowing that it was a bullfinch. And there was a memory of Africa: not Luangwa for once; this was in Tanzania. It was one of the great experiences of the ah-ness of living things.

I was with a sulky, uncooperative driver who had believed that he had wangled himself a morning off because the group of journalists I was travelling with wanted to go shopping. All except one. So the two of us went into Arusha National Park, and he was rather inclined to resent the whole business. No matter, he took me out anyway and left me to work out what I was seeing for myself, which was fine by me. His silence, his refusal to tell me what to look out for, was doubly fine, because otherwise I wouldn't have had this epiphany. The big moment came when the vehicle reached the top of a huge hill and we looked down over a lake: one of those fine sights of open water that always lifts a human heart. Water, after all, is life, and our delight in such watery views is an atavistic as much as an aesthetic response. All round this lake was a fine pale beach: white sand, but with a warm tone to it. Some freak of geology, no doubt, or perhaps a trick of the light. Down we climbed, hairpinning along the forested track, one grumpy, one joyous. And then joy turned into something far less ordinary as I realised that the beach was no beach at all. It was solid flamingos.

Not a beach but a sea of pink: not a pleasing sight but one of the most bewildering things you could wish to see, not something that would attract an artist's brush because no decent artist would dare to take it on: too big, too many, too obvious, too easy to sentimentalise, too hard not to overdramatise. A painting of this lake of flamingos would show only the limitations of human craft and mind and soul.

We made a circuit of the lake, and it was a world of impossible wonder. In terms of vision, anyway. The beautiful sight came with a cacophony honking and bugling and an unavoidably hideous smell. This was not heaven, though it had seemed like it at a distance. Once I was in the middle of them all it was quite clear that we were still in the kingdom Animalia: and perhaps all the better for that. Flamingos love to be together, and there must have been around half a million in this colony. There are six species of flamingos, two in the old world and four in the Americas, though I have seen a Chilean flamingo on a marsh in Suffolk. It was an escapee rather than an Atlantic-crosser, and it took a liking to a nature reserve on the coast. One of the wardens grumbled: "It's not a bird, it's a bloody blancmange. Do you think anyone would notice if I shot it?" The flamingo's originality comes from its bill, which filters mud and silt from particles of food, shrimps and algae. This is the only beak that is designed* to be used upside down.

Flamingos represent better than any other group the glorious accessibility of birds. For all humans, birds are the easy way into wildlife. Not only do birds provide instant

---

* When I say designed, I of course mean "designed".

wonder, they also represent biodiversity on a scale appropriate to human understanding. Birds are diverse *ma non troppo*: there's just about enough diversity to excite cries of wonder, but not enough to have us giving up. To have seen more than 400 species of birds in Britain is reckoned to be the mark of a high-achieving twitcher: there are more than 2,000 species of moths available to a British moth-er, and maybe a quarter of a million worldwide. Hummingbirds, with 300-plus, are baffling, but just about dealable-with; the fact that a flamingo comes in six distinct species (this is argued over, naturally) shows that there are more flamingos than most of us dreamed off, but not too many to halt the processes of thought. It is natural for our sense of natural wonder to start with birds: birds are the portal that exists between humans and the wild world. Birds are our wardrobe. As the four children entered Narnia* through the wardrobe and found a world of magic and talking animals, so the rest of us can look at birds and enter a world more meaningful than our own, one that is not restricted to a single species.

Birds give us diversity on a scale we humans find manageable: they also bring us abundance in a way that we can rejoice in. Recent conservation messages have tended to be about biodiversity, and diversity is essential to life. But life can also be about abundance: about teeming numbers, about the sense of the wild world as a population, a vast social organism, a community that goes on and on and on. The flamingo shore was an epiphany of abundance: of the importance of numbers. The passenger pigeon used to exist in flocks that blackened the sky. They were shot to bits just for the hell of it. After the last one died in 1914, it was speculated that these birds had their being in numbers and could only survive in numbers. A small flock of passenger pigeons was an oxymoron: or rather, it was a fly-past of

---

* In *The Lion, the Witch and the Wardrobe* by CS Lewis.

the living dead. Perhaps the same is true of other species that like to be together in large numbers: in Britain rooks, starlings and gannets. It is not enough to conserve token numbers of each species in the belief that this is what diversity is all about. We also need to look after numbers. As we lose abundance the entire wild world is slipping away from us.

James Bond and the kraken

"Behind him the water quivered. Something was stirring in the depths, something huge... Bond stared down, half hypnotised, into the wavering pools of eye far below. So this was the giant squid, the mythical kraken that could pull ships beneath the waves, the fifty-foot-long monster that battled with whales, that weighed a ton or more... God, the thing was as big as a railway engine!" This is from James Bond's encounter with a giant squid in *Dr No*, in which Bond duels with the great beast armed only with a makeshift spear. I shan't risk spoiling the ending for you.

There's another great battle with a giant squid in Jules Verne's work of 1870, *20,000 Leagues Under the Sea*. The squid attacks the *Nautilus*, the great submarine, and devours a crew member before it's finally beaten off. Verne's work is always balanced on a knife-edge, mixing science and fantasy with reckless creativity. The *Nautilus* visits Atlantis at one stage of its journey: but the giant squid is real. It's just extremely hard to believe in. Easier to believe in Atlantis.

Not least because it's an invertebrate. Go to the enchanted temple of the Natural History Museum in London. In the lobby the mighty Diplodocus stands: a dinosaur more than 100 feet, 30 m, in length. In one of the inner halls a blue whale is suspended above your head: as long, but a great deal bulkier, because Diplodocus is mostly tail. The blue whale is the largest animal that ever existed on the planet. They're both vertebrates: you can count the endless vertebrae in the Diplodocus skeleton if you've a mind to.[*] But the giant squid is a mollusc. Just like the snail in your garden, just like the slugs mating with such disturbing passion earlier in this book. The kraken, the great monster of the deep, is related to cockles and mussels alive, alive-oh.

Small wonder nobody believed in it. The plight of the poor seamen who cried "squid" was worse than that of the boy who cried "wolf". It was years before anyone believed in the giant squid. In 1848 Peter McQuahe[†] claimed the sighting of a sea monster near the Cape of Good Hope. Richard Owen, a big man in science, coiner of the word "dinosaur" and the driving force behind the Natural History Museum, took exception and poured scorn and personal vituperation on McQuahe, implying that he was unworthy to command one of Her Majesty's ships. A few years later a French captain claimed a sighting of the giant squid and even managed to bring a piece back. Scientists laughed and said that it was plant material, though Verne liked it and put a huge squid his book.

Then in 1873 a Newfoundland fisherman named Theophilus Picot didn't exactly catch one. It was more vice versa, or so he claimed. He said that the thing attacked his boat and tried to drag it down. What's certain is that Picot managed to hack off both feeding tentacles and bring them back to shore. Moses Harvey, the local rector, realised that this was something momentous: the 19 foot, 5.7 m tentacles made it clear to the world that the giant squid was not a fisherman's tale. The kraken was real.

Why did it take so long for such a huge animal to make the great shift from mythology to science? It lives in the

---

[*] People regularly used to steal vertebrae when the skeleton was displayed with the tail dragging along the floor. The exhibit is a plaster cast, and the museum used to keep spare vertebrae for use when needed. These days, with dinosaur revisionism in vogue, Diplodocus is displayed in a far more dynamic pose, with the tail whip-lashing high into the air, inaccessible to souvenir-hunters.

[†] See *Kraken: The Curious, Exciting and Slightly Disturbing Science of Squid* by Wendy Williams.

8. COLOSSAL SQUID

depths, thousands of feet down; a squid on the surface is dead or dying. Dead squid at those extreme depths get consumed long before their bodies float to the surface. And even if you managed to get hold of a surfaced or beached squid, it is extremely unwieldy and constantly being eaten. It also keeps falling apart, since it lacks a skeleton as we vertebrates understand the term.

The giant squid is now accepted as a fact of oceanic life. It has been suggested that sperm whales eat a couple of giant squid every week. The mantle – the body bit – is about 6 feet long, a couple of metres; the tentacles extend that to 40 feet, 12 m. They live by catching fish with the two longest tentacles, "like a two-tongued toad", as one scientist put it.

It was believed to be the world's largest invertebrate. But in the past century odd bits of an even bigger squid started turning up, and in the past decade people have found intact specimens. This is the colossal squid: an animal with a mantle twice the size of a mere giant and with tentacles that, when alive and active, take its total length to something like 60 feet, 18 m.

And still very little is known about it. It has eyes that are about 12 inches, 30 cm across, and it is considered to be more formidable than the giant, and to be an apex predator like the great sharks and the toothed whales. Some scientists suggest that the animal is also pretty intelligent: though not necessarily in the sense that "intelligent" means "like us humans". We know little about the colossal squid beyond the fact of its existence. Our own planet remains a place of mystery. "Why aren't we spending more on our oceans?" asks Clyde Roper, the squid scientist. "We know more about the moon's backside than we do about the ocean's bottom."

Instant birder

"I'm not really interested in looking at birds," they say, sometimes apologetically, sometimes almost belligerently. They have come to Africa, come to the Luangwa Valley, to see lions and tigers. They're not birdwatchers, for God's sake. Like anyone else who has acted as a guide in such circumstances, I smile pleasantly, make a remark that commends the person's large-hearted frankness in making such an observation and stresses the total lack of any kind of obligation towards avifauna in general and that of the Luangwa Valley in particular – and bide my time. Then I show them a lilac-breasted roller. Never fails. I have a party of instant birders.

Lilac-breasted rollers like to perch on dead trees, in places of maximum exposure. They are about the size of a small crow. Here's the description from the magisterial *Roberts' Birds of Southern Africa*: "Crown light green, back light brown, rump blue, tail blue, forked... breast lilac, belly blue; in flight wings bright blue." In other words this bird is a great feast of the most extraordinary and vibrant colours. When they are in an extravagant mood, which is often, they fly upwards, croaking rhythmically, and then explode into a series of cackles, diving headlong towards the ground, pulling out in the nick of time to loop the loop before making a low-level pass with a comic side-to-side rocking that shows off the electric blue of the wings. Here is an explosion of colour, of extravagance, of over-the-topness. Here is the bird that makes birders of us all.

After that, we visit the bee-eater colonies that dot the

banks of the Luangwa: soft cliffs stippled with rhythmic holes that seem to have been blasted out by a giant shotgun, in and out of which the most astonishing birds zoom in eye-baffling numbers. When alarmed, they take to the air hundreds at a time. The carmine bee-eater is another confection of almost absurd extravagance, rose red picked out with blue and green: an acid-trip fantasy of a bird that comes at you in hordes.

But perhaps that's what you expect of gaudy foreign birds. Perhaps you think that birds of Britain can't begin to compete with such creatures. Well, walk with me across the marsh behind my house and sit with me at my favourite sitting place, and be still and quiet for a while, Often enough, the moment happens. You know when you wave a sparkler around in the dark? It leaves a burning line hanging in the air after it has gone. A kingfisher seems to do much the same thing: a dead straight line of laser blue smoking low to the water. I'd happily take a quid for everyone who has ever told me "I'd love to see a kingfisher". They're there waiting for you, and if you take up the habit of sitting still in wet places, the chances of you seeing one eventually are around 100 per cent.

Here are three more groups of portal birds, of wardrobe birds, of birds that inspire humans to admiration and wonder and fill us with a lust to know more about the world we live in and the kingdom to which we all belong. And they are closely related: bee-eaters, rollers and kingfishers are all brightly coloured birds that sit and wait for food to come past, often in prominent exposed positions. There is a book* devoted to them and it lists and illustrates 123 species, each one of them an impossibility, an absurd polychrome extravagance, an almost gratuitous demonstration of nature's power to do just whatsoever it bloody well likes. Evolution can make a sparrow, it can make a Bohm's

---

* *Kingfishers, Bee-eaters and Rollers* by C Hilary Fry, Kathie Fry and Alan Harris.

bee-eater, a purple roller and a stork-billed kingfisher.

There has to be a moment when wow turns to why. Darwin's nausea at the peacock's tail was a malaise he eventually conquered: at first sight – though these are birds that seem to demand repeated and prolonged gazing – there seems no reason whatsoever for their 123-times repeated extravagance. It doesn't work as camouflage: quite the contrary. And it doesn't work as an aid to sexual selection like the peacock's tail: in this group, the females are every bit as glamorous as the males. They're all gorgeous.

The best answer is intraspecies recognition. We are able to recognise a fellow human from one of our nearest relatives with a nanosecond glance: you don't often find people saying *phwoar* – oh no, that's a chimpanzee, isn't it? It's important for all species to know themselves and the community they belong to. Ultimately that's because copulating with the wrong species is a dead end, but long before you reach such a point, a recognition of your own species – as rivals, as allies, as fellow flock members, as relatives, as potential or actual partners – has to be a crucial aspect of the way any species – certainly a warm-blooded vertebrate – understands the world. A dog knows it's a dog even when surrounded by humans, and will greet a fellow dog and a human differently. In the Luangwa Valley, white-fronted bee-eaters and little bee-eaters live alongside carmine bee-eaters but they don't get mixed up. They couldn't. The explosion of colours has a biological and an evolutionary function.

But do they really have to be so damn beautiful? Science is silent on that point. We have considered the perfection of tapeworms in this book: we are, then, entitled to pause a moment's gazing on the lilac-breasted roller. *Che bello! Che bellissimo!* Perhaps not all things, but certainly *some* things, really are bright and beautiful.

Superslug

Tentacles, large eyes that look nearly but creepily not quite human and a considerable intelligence… standard recipe for a science-fiction alien, and a pretty good fit for an octopus. A member of the Ood in Dr Who is more or less an octopus in a suit. Perhaps even more than a giant squid, an octopus is an unlikely relative of the snails that slide across your garden wall and the slugs that slime towards your lettuces (when not engaged in a spectacular bout of sex).

But octopuses are molluscs all right, even though they have taken a dramatically divergent evolutionary route from snails.[*] One theory is that the loss of their shell gave them much greater mobility and flexibility of behaviour, and in order to make this strategy work in a changing world, they evolved intelligence. Molluscs and vertebrates spilt off from each other millions of years back: this is intelligence as a piece of convergent evolution. Just as a bat and a bumble-bee can both fly, so a human and an octopus can both think. But just as bats and bumblebees don't fly the same way or with the same sort of mechanism, presumably humans and octopuses don't think the same way.

Plenty of complex and meaningful experiments[†] have

---

[*] Octopuses is by far the best plural, as with platypus. Octopi is plain wrong, because octopus is Greek, not Latin and octopodes is linguistically correct but insufferably pedantic.

[†] The issue of octopus intelligence comes up in *Sweet Thursday*, John Steinbeck's sequel to *Cannery Row*, albeit with an unacceptable plural. Doc gets distracted mid-seduction: "'Octopi are timid creatures really,' said Doc excitedly. 'Most complicated. I'll show you when I get them in the aquarium. Of course there can't be any likeness, but they do have some traits that seem to be almost human. Mostly they hide and avoid trouble but I've seen one deliberately murder another. They

been done with octopuses in mazes and in other problem-solving situations, but the anecdotal evidence has a vividness that can never be found in the columns of figures required by scientific rigour. Octopuses are the only inverts known to use tools: they have been observed manipulating coconut shells to create a safe hiding place. They are the only invert known to indulge in play: an octopus was left alone in an aquarium with nothing interesting happening, but was supplied with a plastic medicine bottle. It contrived a game in which it fired the bottle across the tank so that it was sent back by the aeration system; it did this 20 times before getting bored. Octopuses are always climbing out of tanks, sometimes to invade neighbouring tanks in search of food or excitement or sex. They hide when they know they are required to take part in onerous experiments. They demonstrate memory, short and long term. In one experiment, two identically dressed people had daily dealings with the same octopus: one gave it food, the other irritated it with a bristly brush. After a few days it hid from the one and flaunted itself before the other. Octopus experimenters are clear, in an informal way, that octopuses have different personalities: some inquisitive, some cautious, some impetuous.

It's clear, then, that we have to adjust our minds. We have all looked into the eyes of a dog and wondered just how much of a mind the two of us have in common. But an octopus is a mollusc: a creature dismayingly remote from all of us vertebrates, separated by a great deal more than such trifling taxonomic considerations as order. We are talking phylum-deep differences here: an octopus is kin not to us or our dogs but to a plate of winkles. Here is the slug that encountered krypton: one of the humbler creatures of the earth that apparently got contaminated by some alien

---

appear to feel terror too, and rage. They change colour when they're disturbed and angry, almost like the rage blush of a man.' 'Very interesting,' said the girl and tucked her skirt in around her knees."

substance and came back not only different from its ancestors but disturbingly cleverer. As humans are to other vertebrates, so octopuses are to other invertebrates:[*] cleverer to a disturbing, to an almost unnatural degree. This is so much the case that octopuses are regarded by law as honorary vertebrates. The laws that govern experimentation on animals[†] allow you a pretty free rein on all invertebrates, apart from cephalopods, the group that includes octopuses.

There are around 300 species of octopus, and the way they live reflects their intelligence and versatility. They are adept at catching small animals and escaping from large ones. The octopuses that live on the sea floor mostly eat crabs, polychaete worms[‡] and some of their fellow molluscs like whelks and clams; those that live in the open oceans eat shrimps and fish. Their weapon is a beak hidden within the crown of tentacles, which can impart an injection of poisonous saliva that paralyses. The poison of the blue-ringed octopus is fatal to humans, though outside fiction humans are not the preferred prey of octopuses. The beak is also used for dismembering prey. They have been known to climb on board ships to eat crabs. Octopuses famously eject ink and escape from predators under its cover: the stuff hangs in the water and confuses not only eyesight but scent. They can use speed, by means of jet propulsion, squirting water to shoot off headfirst, tentacles trailing. They are adept at finding and creating hiding places. They can also use camouflage to hide in plain sight. Some species can put up a performance to startle predators,[§] sometimes changing colour dramatically, sometimes using mimicry to imitate dangerous creatures like sea-snakes, lionfish or eels. They are unusually short-lived for animals with such

---

[*] Just a reminder here that invertebrates are not a single coherent group, but many different groups.
[†] The Animals (Scientific Procedures) Act of 1986.
[‡] Polychaete worms are sometimes called bristle worms and belong to the phylum of annelid worms already discussed in these pages.
[§] Technically a deimatic display.

intelligence: six years at most. The male lives only a few months after mating; the female dies as soon as her eggs have hatched: she looks after the unhatched eggs without eating, and when that job is done she has nothing left to do.

Dr Who has two hearts, but that's not going to impress an octopus, which has three. It has a highly complex nervous system, as you'd expect, and its intelligence is only partly located in its brain. A degree of intelligence lies in the limbs themselves, which, truly disturbingly, have limited autonomy. Most species are fairly small, but here are a few whoppers: the giant Pacific octopus has a leg-span of 4.3 m, 14 feet, and a record weight of 71 kilos, 156 pounds.

Squids and octopuses make up the class of cephalopods, just as we belong to the class of mammals. They possess two qualities that are very unusual among invertebrates: size and intelligence. It is troubling enough to think of seriously sizable inverts in a form so different from us vertebrates: but it is far more disturbing to think that intelligence is something we share with creatures so distant from us in evolutionary terms. Octopuses comprise 300 species as remote from us as the garden slug, and yet are capable of thinking for themselves, making a hide-out, solving problems, and pursuing their lives in a bright, active and ingenious way. You can't say that because an animal is intelligent it must be a mammal, and therefore comfortingly close to us humans. Intelligence can exist independent not only of humans, but of our close relations. Real intelligence exists way outside our own kind.

The Clever Club

In one of Aesop's fables a thirsty crow funds a jug of water. But the water lies too deep for him to reach, and the jug is too big for him to push over. So the crow drops pebbles into the jug until the water level rises high enough for him to drink: a tale that extols the virtues of thought over brute strength, a hymn to resourcefulness, persistence and intelligence. The surprising thing about the fable is that it stands up to rigorous scientific verification. Crows really are capable of solving that problem. I've seen them at it. And it's a performance that, if we dare to think about it, will muddle us up horribly and force us to look at the world in a dismayingly different way.

Night and day. Good and evil. Men and women. We have a deep desire to keep such things separate: in adamantine caskets, the twain never meeting. Sacred and secular. War and peace. Work and play. Tragedy and comedy. Life and death. But life, if you are forced to look at the stuff at all closely, is forever blurring the boundaries and swapping our certainties for questions. Perhaps death does too, in an equally annoying way: a prolonged terminal coma is, after all, not exactly life.

I made a trip to Cambridge, or to be specific, to Madingley, a village close by, to listen to a series of dancing dizzying concepts that can't help but dissolve these adamantine caskets. Here at a Cambridge University lab, you can find Nicky Clayton, Professor of Comparative Cognition. She is by temperament a dweller in the spaces between the hard-and-fast. She had invited me to visit her birds. Her crows.

Together, they were working on the shared ground between the human and avian mind. So there she is, plenty blonde and about five foot four in monster heels, calling the jays in a cooing soprano: "Hello, boys! Helloooo!"

It was cheerful and soppy and slightly embarrassing; I'm just the same when I'm talking to my horses. With these birds, Nicky has compiled a series of elegant (what other kind would she do?) experiments demonstrating that when it comes to intelligence, crows are right up there with the great apes. Broadly speaking, if a chimp can do it, so can a jay. The water-jug fable has been re-enacted in this lab: you don't make the crow thirsty to the point of desperation, which would be unethical; instead, you float a waxworm, a prized morsel, on the water, and leave the jay wondering how to reach the damn thing. The use of pebbles followed. Give them painted pebbles and identical-looking Styrofoam fakes, and they will ignore the fakes, and use only the proper worm-bringing stones. Nicky talked about "the Clever Club": animals that have something we humans are forced to recognise as intellectual capacity. Example: an ability to plan for the future in a thought-out rather than instinctive way. The Clever Club includes elephants, dolphins, the great apes and members of the crow family. And us.[*]

But there's a small difference to consider before we move on. The more intelligent the animal, the more guilty we feel about its wanton destruction. People who will eat a fish with great appetite often feel differently about the idea of eating a dolphin, at least in western cultures. We are disturbed by the haphazard killing of elephants, and even more of chimpanzees and gorillas: here killing becomes uncomfortably close to murder. And yet crows, just as clever, are shot as vermin, and old-fashioned types still hang out their corpses as an awful warning to other members of the crow family. Since visiting Nicky's crows I have taken to watching rooks

---

[*] If you want to draw a line between humans and our fellow apes.

in a very different way: within any given flock, any one rook is likely to associate closely with one other rook. They believe in a long-lasting pair-bond even though they are a flocking species. Carrion crows, magpies and jays are more usually seen as half of a pair than not: the old magpie adage of one for sorrow, two for joy gets its impetus from the fact that magpies normally come in pairs, which makes them far more likely to be joy-bringers than not.

At one stage, we believed that humans could think and animals couldn't, and that was both the difference and the end of the matter.* But it wasn't. As we have seen, every time philosophers redefine the word "think", animals and their associated researchers come up with more problems: signing chimpanzees, problem-solving dolphins, elephants with an understanding of death. It's like an arms race. We have grudgingly accepted a few more mammals into the Clever Club that used only to accept humans. But Nicky's research has demonstrated beyond all doubt that some birds have a right to membership as well. Big brain relative to body size, long life, extended development period spent with kin, allowing social learning: these are things all members of the Clever Club have in common. It's not a hard thing to grasp: after all, we fit right in.

I was delighted to learn that one of Nicky's PhD students is researching the question of whether males know what females want. Jays go in for long-term relationships, and there are many ways in which the bond is nurtured and strengthened throughout its existence. One of these is gift-giving. Do male jays understand female preferences and make gifts of the food they like best, just as I gave my wife a cup of tea rather than a pint of lager this morning? It's the fuzzying-up of yet another barrier.

Nicky's conversation dances thrillingly around all these topics, flip-flopping effortlessly between things that are not

---

* *Cogito ergo sum* and all that.

supposed to be flip-floppable. But then Nicky is a dancer, and scientific adviser to the Ballet Rambert. Art and science are not separate things for her: just one more opportunity to flip-flop. Recent decades have given us some of the finest science writers in history. Stephen Jay Gould and Edward O Wilson have both written books seeking a union between science and art, and the brilliance of their writing puts that principle into action. No problem, then, with throwing another art form into the mix. Nicky dances 15 hours a week, her preferred form being Argentinian tango, though she flip-flops into salsa. Dance quite clearly informs her life and her thought and her work. Perhaps we should simply regard science and art as one activity: humans seeking to understand the world.

What do I mean, humans? Jays solve problems and plan for the future, while 50 yards from where I write, I have watched horses express themselves in movement, dance, if you like, and heard song thrushes perform their own intensely individual compositions. Human thought, human achievement, human aspirations: we talk about these things as if there couldn't possibly be any other kind of thought, achievement, aspiration. It strikes me that we're all in it together.

Nautilus but nice

Skip this chapter. I really should be doing the same myself: after all, I've got a book to finish. I should be hurrying along to other molluscs, instead of lingering with the cephalopods. But I find it impossible to continue without spending a moment with the nautilus. I'm not really sure why: it's not as if I've ever seen one. But for some reason they have a deep and elusive meaning for me.

These creatures have always had something magical about them, at least for me. And it's complicated. There's a distant memory of watching David Attenborough explaining the nautilus shell, which is used as a bailer by some island people around Papua New Guinea. I have half-remembered visions of seeing pictures of these strange things in the sea: coiled shells from which a wreath of tentacles tentatively emerges. Then there is the name: the *Nautilus* was Captain Nemo's impossible submarine in *20,000 Leagues Under the Sea*; that was a book to dream on all right. But I think more than anything else it's the fact that these curious creatures bring me back to illustrations I saw in my boyhood: illustrations of the world beneath the seas of the Devonian era. Lost worlds and extinct animals had an eerie fascination for me, as they do for most children; most adults, too, I suspect.

The nautilus group is ancient enough, 500 million years or so. But what matters more to me is that they look a great deal like ammonites. Ammonites are perhaps the definitive, the archetypal fossil: a beautifully coiled, ridged spiral that makes an almost abstract pattern of life. There is one on my desk; the tentacles are there at the shell's mouth in blurred outline. It's not a great fossil, as collectors rate them, but it's on my desk and that gives it meaning. In truth, ammonites are more closely related to present-day squids and octopuses than they are to the nautilus. But they look like ammonites, and they look as if they have just swum out of the pages of *The Golden Treasury of Natural History*,* by Bertha Morris Parker, a book full of wonders that my grandparents gave me for Christmas when I was, I think, nine, a book that shaped my understanding of the world.

The magic of the nautilus lies in this similarity to the perfect fossil. It reminds me of those artists' reconstructions of ancient seas: trilobites crawling about on the ocean floor, and above them, fleets of ammonites, tentacled garlands protruding from the elegant spirals of their shells, improbably propelling themselves about the oceans of the lost world.

The nautilus species that survive today are from a very ancient line, and there are very few species left: just half a dozen. The chambered nautilus is the best known: the creature advances through its shell throughout its life, vacating an old chamber and building a new one, and as it does so, creating a work of beauty and mystery as well as an ad-hoc bailer. It can have as many as 90 tentacles in two circles on its head, and can measure up to 27 cm, 11 inches, in diameter. It can close its shell with a leathery hood formed from two tentacles, and it swims by jet propulsion, like an octopus. It can adjust buoyancy by adding and subtracting water from its shell. Unlike octopuses, nautiluses have no claims to higher intelligence, though they have been shown to possess rudimentary long- and short-term memory: perhaps the seeds of intelligence are there, but were simply unnecessary for the shell-dwelling lifestyle.

---

* Look on this book as my attempt at a grown-up (fairly grown-up, anyway) version of *The Golden Treasury*.

The notion of a "living fossil" is a confusing one. It implies that the creature so-named should be extinct, and is some kind of failure. It is the very opposite, in fact: an example of a design that is so effective it has functioned pretty well unchanged across millions of years, escaping major extinction events as it did so. The nautilus gives us a vision of ancientness, for these creatures have survived since the late Cambrian period, products of the great Cambrian explosion that changed life on this planet forever. Time past and time present are both perhaps there in the shell of the nautilus. And the ammonite on my desk looked very similar as it cruised and tentacled its way through the ancient oceans. "It looks nice on your desk," my older boy said. "Makes you look like a real naturalist."

Bell-beat of their wings

I am writing these words in my usual workplace, which is a hut that looks over a fragment of Norfolk marsh. You'll understand, then, that it's wonderfully filled with potential distractions. There is no momentary break that can't be extended by a long stare at the uncompromising landscape before me. Often, I am released from my computer screen by the sound of whiffling. They were at it as I walked towards the hut to start work this morning: four whifflers in a hard, straining line low in the sky.

Mute swans don't get a great press from bird-writers. They – perhaps I should say we – prefer their more glamorous and much wilder relations, the whooper and the Bewick's swans. These are birds that fly in from the Arctic to spend the winter in Britain: birds with no history of taming, birds that were never farmed to make up a medieval banquet, birds that never queued up for a child's bag of crusts. But there is an epic quality about a mute swan that's best appreciated when not looking down at them from the edge of a duck-pond. Mute swans are better when you look up.

They are the world's second-heaviest flying bird. The all-comers' record is held by the kori bustard, a chunky bastard of a bird that I have come across often enough in Africa. To see two or three take to the air is like watching the evacuation of Saigon.* But mute swans are not far behind them

---

* A kori bustard weighs in at 41 lb, 19 kg, a mute swan at 39 lb, 18 kg. In bronze medal position is the Andean condor at 33 lb, 15 kg.

and as birds go, they are almost as homely as robins. You can hear the power, hear the effort needed: no mute swan pretends that flying is an easy matter. The sound of the air being hammered* through the stout, almost unbreakable quills fills your ears as you stand beneath: whuff-whuff-whuff, beat by effortful beat, neck extended like the arm of a novice skater desperate for the safety of the rinkside. See them land on a lake: they come in at a shallow angle, lower the enormous black webbed undercarriage and ski along the surface of the water till they have decelerated enough to sit down. See them take off: they must run along the surface, flapping madly, leaving a great runway trail of double puddles, looking horribly unserene and undignified, until they have enough speed to take whifflingly to the air. If they are forced to crash-land in a confined space like a suburban back garden, they can't take off; many a bemused householder has called in the RSPCA to help the bird find its wings again. Flying is a big deal for a mute swan: but it's no luxury, no mere bonus, nor is it something they only do at moments of desperation, like a pheasant. Flying is at the heart of their existence.

That is true of practically all waterbirds; I use the term here to cover ducks, geese and swans.† They can swim, yes, and that's great, but their existence is made possible and defined by their ability to fly. Water, even before food, means safety. On water, a bird is safe from any land predator. But it needs to fly there and fly off again.‡ Some ducks§

---

* *All's changed since I, hearing at twilight,*
  The first time on this shore,
  The bell-beat of their wings above my head,
  Trod with a lighter tread.
  WB Yeats, "Swans at Coole"
† I prefer the term waterbird to waterfowl: waterfowling is a shooting term. To be technical, what I'm writing about here is the order Anseriformes.
‡ But there are three species of flightless steamer ducks; they are all noted for their aggressive behaviour and their fearless attitude to potential predators.
§ The ducks in *Finnegans Wake*, being educated French ducks, say: "Quoiquoiquoiquoiquoi", perhaps in bewilderment at the book they find themselves quacking in, perhaps in bewilderment at the unending forms of life. The

are particularly adroit at this: watch the near-vertical take-off of a group of mallards. The collective noun "a spring of teal" expresses their ability to swap water for sky in an instant. Waterbirds seldom restrict themselves to one patch of water: they need to commute from one to another, in the course of the day, or of a year, for the various different demands of their strenuous lives: feeding, roosting, breeding, wintering. In winter, open water in Britain is full of water-birds, many of them visitors who have escaped from the icier regions where water becomes too inhospitably solid for them. Just a short distance from my hut is a patch of open water where I can see 500 wintering wigeons, ducks that graze on the land when hungry but rest up in the middle of the black waters of the lake, occasionally whistling to each other in far-carrying sibilants. In the summer the waters are relatively open: in winter, it's as crowded in places as the rush-hour Tube.

These waterbirds are defined by the structure of their bills and tongues, which work together like a suction pump, allowing water to be taken in at the tip and expelled at the sides, with chunks of food saved in filter plates called lamellae. But many waterbirds have evolved to take food from outside the water: geese and some ducks graze onshore. One of the great sights in British birding is the dawn flight of pink-footed geese in the Norfolk winter, flying in from the shallows of the Wash to graze on sugar-beet fields during the day: hundreds and hundreds of them in great straggling calligraphed Vs. Some ducks have adapted the basic model and become fish-eaters: the sawbills have serrated bills as their name suggests and they are adept divers and graspers. In this country we have the scruffy red-breasted merganser and the elegant goosander, and occasional winter visits from the impossibly dapper smew.

---

seagulls, however, say: "Three quarks for Muster Mark", thereby giving a name to the elementary particle.

The feathers of ducks, of all waterbirds, are famously, even proverbially, watertight and waterproof. Beneath the shiny top layer of feathers lies a layer of superb insulating down. Water would destroy it, but water never reaches it, so effective is the waterproofing: and this explains why ducks can sit all night on water that would kill you and me within an hour or two. The top feathers are oiled regularly with a substance secreted from the preen gland. Feather maintenance is vital for all flying birds, for without immaculately looked-after feathers the mechanisms of flight break down. A waterbird needs to fuss and preen even more than most, because it must keep the water out or die, just as it must fly or die.

The point of waterbirds is not that they can swim, but that they can swim as well as fly. Like most birds they have their being in mobility, and their ability to relocate, and to escape from dangers that come from weather and from predators. Feathers give the birds flight, and flight gives them areas of the world that few other large creatures can exploit. Flight has opened world after world to birds: and water is just one of them.

She sells seashells

I was still wondering what the hell had possessed me to come. Sunnyhill Junior School in Streatham was bad enough in term-time: what was I doing with the school in the holidays? Why had I decided that I wanted to go on School Journey? I must have been trying to fit in, never a good reason to do anything. After a long train ride, most of which I spent mooning out of the window, we checked into a lodging house. We dumped our bags and were allowed out. We were at Paignton, in Devon, right on the seafront. And as we hit the beach, two simultaneous waves of delight hit the male part of the party. Did I whoop too? Certainly I can remember the unbelieving joy at this instant discovery. The beach was full of round flat stones that were idea for skimming!

Not that I gave a toss. It was everyone else who got excited about the stones. I would never for an instant deny the keenness of the pleasure to be gained from skimming stones across the sea, but it pales into nothing when compared with the joys of the wild world. And Paignton beach was covered with shells.

Shells! That stroke of genius from the molluscs, that hard and lovely thing that endures for hundreds, even for thousands of years after the soft creature itself has died. With my hands scooped together I could pick up dozens of them, generations of them. Tower shells and top shells and periwinkles and whelks and mussels and oysters and razor shells: a wild museum, almost imploring me to fill my pockets and my suitcase with these gritty, lovely trophies:

part of the living world I could pick up and revel in and do so without hurting, still less killing, a single creature. Shells touched my heart with a rare sharpness. I could gaze on them and touch them and wonder at them and look them up in books and – glorious, glorious thing – name them. What made all that even richer was the fact that I could actually own them. I could take them into my home: make them a part of the museum I had established in my bedroom on tottering brick-supported shelves, the letters of their magical names written in maladroit letters. There are shells in *Ulysses*: "And now his strongroom for the gold. Stephen's embarrassed hand moved over the shells heaped in the cold stone mortar: whelks and money, cowries and leopard shells: and this, whorled as an emir's turban, and this, the scallop of Saint James.* An old pilgrim's hoard, dead treasure, hollow shells."

There's always been a mystery about seashells for humans: a sense of value and meaning as Joyce implies. Cowry shells were used as currency in Africa for centuries; they were also used for money in North America. Cowries have been traded as jewellery and as charms, and they have been used for divination. Shells have always mattered to humans, but to a shell-wearing mollusc a shell is life, the universe and everything: the meaning of existence is shell. The shell supports, encloses and protects the soft life within. It is a calcareous exoskeleton, but unlike the skeletons of us vertebrates, shells contain no cells or DNA. Every shell is a small miracle: an inanimate object produced by an animate being. The hard shell of calcium carbonate is not an exclusively molluscan invention: corals, crabs, barnacles (which are not molluscs),† seas urchins, lampshells and bryozoans also go in for shells. Molluscs have done it more times and more thoroughly than any other group: molluscs are the

---

* A scallop shell was traditionally carried by pilgrims to the shrine of Santiago – St James – in Spain.
† Darwin had a special relationship with barnacles, which we'll examine later.

most diverse animal in the oceans, and make up 600 families in all. Most are tiny, but naturally, we will look at some of the whoppers as well. There are shell-wearing molluscs in fresh water, as every aquarium-keeper knows, and plenty on land, as everyone who has ever sat in a garden knows.

The shell is, as it were, willed into being. To be more scientific, it is secreted by specialised tissue called the mantle, something every mollusc, shell-wearing or not, possesses. The secretion contains a protein called conchiolin. Every shell, whether or not it contains a living animal, looks like an entirely finished thing, but it grows throughout the life of its creator, a process that begins with the addition of a fresh band of conchiolin along the existing edges of the shell. Internal layers can also be added to make the shell thicker. There are about 100,000 known species of shell-wearing molluscs; there is some dispute as to whether or not these should be classified separately from shell-less molluscs.

We humans have an intimacy with shelled molluscs: a special relationship. Shells have been used as money, as Joyce writes in the fragment quoted above. A few years ago, archaeologists in Morocco discovered shells that had been perforated, presumably as human adornments. The perforations themselves were 80,000 years old. Were they worn by males or females? Did they wear them to demonstrate their status? Or to make themselves more beautiful? No doubt the answer is the same as it is for the jewellery in our shops now: both. "Stephen's hand, free again, went back to the hollow shells. Symbols too of beauty and of power. A lump in my pocket. Symbols soiled by greed and misery."

22:1

We set off from Stonehaven, a tough-looking harbour on the east coast of Scotland, just south of Aberdeen. The sea fret had come down like the curtain at Covent Garden: you could see maybe 20 or 30 yards. So we went out in an open boat looking for dolphins. Not that we'd be able to spot any, but there was always the possibility that the dolphins would spot us and cruise over to check us out, dolphins being full of interest in what is going on. They didn't, but it was an unforgettable trip. For much of it we were out of sight of land. For much of it, we were out of sight of practically everything. It was as if we were in the middle of a small bowl of light, and the rest of the world was a soft thing of pearl and silver – and into this bowl and out again flew bird after bird. Now it is a thing that always fascinates me about the sea: as soon as you are even a short distance offshore, you get a completely different set of birds. As soon as you let go of the land, you find the birds that have done the same thing, but on a career basis. And so, in and out of our bowl of light, there cruised first gannets, the spear-billed fishers who plunge from impossible heights into the sea, with sharp-pointed wings 6 foot, damn near 2 m, in span. Wheeling and turning with wings that seem far too long for their bodies, the shearwaters. Every now and then the Dracula silhouette of skua. And then, whirring into sight and out again, with wings that look far too small to support such a chunky little bird, the auks: guillemot, black guillemot, razorbill, puffin: flying with all the grace of a bath-toy powered by a rubber band. These birds are at their best swimming beneath the sea after fish. But like ducks, geese and swans, seabirds are able to exploit the sea so effectively because they can fly.

These birds of the sea have as little to do with the land as possible. Land is only useful to them as a place to breed: if they could build a nest and lay an egg on the surface of the sea, like the halcyon, they would certainly do so.* They come to land reluctantly, and for as short a time as possible: and it's not what they're good at. A cliff ledge a couple of inches wide with the restless sea a hundred feet below: that's as far away as a guillemot ever wishes to get from the ocean. These seabirds are creatures of the deep sea, and when not involved in making more gannets or guillemots, they spend all their time out on the sea, feeding beneath it when hungry and roosting on its ever-shifting surfaces when not.

The birds' power of flight has given them not just lakes and rivers and inland waters but the entire ocean as well: all that three-quarters of the earth's surface that isn't covered by land is accessible to the birds, and most of those that exploit it do so by combining the ability to fly with their skills as swimmers. Flight gives them mobility, the ability to move from one food source to the next, and of course, the ability to reach a suitably inaccessible nesting site: most seabirds like rocky places, precipitous cliffs that beetle o'er their base into the sea, preferably on a flyspeck island that no human ever visits. Just as some mammals have adapted to take advantage of the riches of sea, so have many species of birds. Like seals, they are tied to the land for breeding, but their power of flight gives them great versatility and a wider choice of breeding ground. The five species of frigate birds have the largest wingspan relative to body weight of all birds. They can't swim, they can't take off from a flat

---

* The halcyon is a mythical bird that was able to calm the wind and the waves in order to nest on the surface of the sea, hence of course halcyon days. A genus of kingfishers bears the name *Halcyon*.

surface and they can hardly walk. That means they have to roost on land rather than on the surface of the sea, and they do so on cliffs and trees, where they also breed.

The masters of this ocean/airways double are the albatrosses, 21 species of them, with the wandering albatross supreme among them with a wingspan of nearly 12 feet, more than 3.5 m. They are the champions of glide: no bird uses the wind more efficiently; they can cover 1,000 km, more than 600 miles, a day without so much as a flap. They use two techniques: dynamic soaring and slope soaring. In the first, they gain height by turning into the wind, rising until they are right on the point of stalling out of the sky, and then they regain speed by turning downwind. They can carry on with this rising and falling perpetually, or as long as the wind supplies them with energy. In slope soaring, they exploit the updrafts that are created as the wind hits the big waves and is deflected upwards. The wings of an albatross are equipped with a shoulder lock, which means they can keep their wings outstretched with no muscular effort: their heart rate when gliding is not much higher than their resting heart rate. They have a glide ratio of 22:1 or better: that is to say, they can gain 22 m horizontally for every one dropped vertically, which is staggeringly efficient. They manage this with the help of the tubes in their beaks – they come from an order called Procellariiformes or tube-noses – which measure their airspeed very accurately indeed, like the pitot tubes on an aircraft.

There is naturally a payback: they aren't much good at powered flight and calm days leave them grounded, or rather, sitting on the surface of the sea. But still days are as rare as albatross teeth in the Southern Ocean, where most albatrosses live, and in the northern Pacific. They aren't found around the equator, where the doldrums and the frequency of dead-calm days would make their lives impossible. There is an exception: the waved albatross breeds on the Galapagos on the equator and uses the winds that are caused locally by the cold Humboldt Current. Albatrosses are phenomenally long-lived birds: the oldest living bird, as I write, is a Laysan albatross that was first ringed on Midway Island in 1956. Of the 21 species, 19 are classified as threatened.

Fearful the death of the diver must be

There are five orders in the phylum of molluscs. We've met the cephalopods, which comprise the octopuses and squids and nautiluses; there are the minor orders of chitons and tusk shells, and the rest are divided into two big ones. There are the gastropods, which we'll come to in a moment, and the bivalves. Bivalves have hinged shells, shells like a game of two halves: two doors, the meaning of the name. Oysters, mussels and clams and many others: this is a highly successful design, and there are getting on for 10,000 known species. The larva produces a single uncalcified shell, which is then enveloped by two mantles, the shell-creating tissues of the molluscs. Each mantle then becomes a separate centre for calcification, each making a shell that is the mirror image of the other. You can see this with asymmetrical oysters: if the two shells weren't equally asymmetrical they wouldn't make a pair, wouldn't close and wouldn't be any use.

The most celebrated bivalve is the giant clam, which is famous for grasping the hands or feet of divers and holding them helpless until they drown: a classic example of the hostility of nature to humankind. We live in a world in which you can't even trust a shellfish not to kill you. There are versions of the US Navy Diving Manual that tell you how to escape from the clutches of a giant clam; you do so by cutting the living creature from inside the twin shells. All this is fascinating because it isn't true. Not even a bit. Giant clams don't attack and drown humans, even inadvertently.

True, they are pretty big, and true, they open and close, but they are not hostile. They close as a defensive, not an aggressive, measure, and they do it so slowly that you put your hand in and out several times over while they are doing so. And even when they have closed, there is usually plenty of room to slide a hand or even a foot out again: they don't close tight like a gin trap. The giant clam as the static slayer of the deep is a human fantasy about the hostility of the wild world, that's all. I'm reminded of a poem, set to music, that was a great favourite of Uncle Matthew in Nancy Mitford's *The Pursuit of Love*:

Fearful the death of the diver must be,
Sleeping alone in the de-he-he-he-epths of the sea.*

Giant clams are spectacular creatures despite this disappointing lack of lethal tendencies. They do something much cleverer than killing humans. They have become farmers. But first the size of them: the giant clam fantasy wants them to be at least as big as Volkswagens, so the exact dimensions may come as a bit of a disappointment. All the same, they're pretty big, certainly when compared with the shellfish that turn up on plates, and they get big relatively quickly because of their skills as farmers. A respectable size for a giant clam is around the 4-foot mark or 120 cm across and weighing in at 440 pounds, 200 kg. Only the shells are left of the largest specimen ever found. It's now in a museum in Northern Ireland and measures 4 feet 8 inches, 137 cm, across and weighs 510 pounds, 230 kg. Add another 20 kg for the animal itself and you have a seriously impressive creature.

Giant clams are found in warm shallow seas to a depth of about 20 m, 66 feet, in the Pacific and Indian Oceans, South

---

* Misquoted as "walking alone" in the novel; the poem is by G Douglas Thompson, set to music by Edward James Loder.

China Sea and off the Philippines, and they can live to be 100. They are scarce and have gone extinct in places where they were once common, for they have been overexploited for food. The truth is that giant clams need to beware of divers. Their preference for shallow seas makes them accessible to humans. They live in shallow water because they need the sunlight, not for themselves but for the algae they farm. The giant clam's evolutionary breakthrough is a symbiotic relationship with single-celled algae that it grows in the tissues of its mantle. The metabolic products of the algae are added to the food the clam takes in by filter-feeding, which means that the animal is well supplied with nutrients even in nutrient-poor water – so it can grow fast and to a splendid size. Once it has gone through a free-swimming larval phase, it settles down and can't move at all, which makes sex a little difficult, at least if you think in human terms. With clam sex, both partners lie back and think of whatever it is that inspires a clam to sexual performance. They are hermaphrodites but they can't self-fertilise. They release both sperm and eggs and neighbouring clams are alerted to this by a trigger substance that readies them for reception. A little like telephone sex, then.

No flying, please, we're birds

Flight is the honour and glory of a bird. So what is the first thing that a bird does when given the opportunity? It gives up flight. New Zealand went through 20 million years without mammalian predators, and so the birds started to become mammals. Or to be more accurate, they filled ecological niches that are normally filled by mammals: a kiwi has feathers that look fur, it lives on the ground, it can't fly, it uses scent to track down invertebrate prey, and unique among birds, its nostrils are at the end of its beak. New Zealand is the world centre for flightless birds, with getting on for 20 species. These includes the kakapo, a parrot that has become a forest browser, the takahe, a rail that looks like a chunky moorhen and eats grass, and wrens that creep across the forest floor like feathered mice. New Zealand has five species of kiwi and six of penguin.

Flightlessness is not an evolutionary or taxonomic category. Species from many groups have shed flight and evolved flightlessness. All it needs is the right circumstances. The obvious explanation is that flight, being highly expensive in terms of energy, is dropped as soon it becomes a luxury rather than a necessity. It has also been suggested that flightlessness is a survival adaptation: you are less likely to get blown off your island home into the limitless ocean if you don't take to the air. It is also obvious that if circumstances change, a flightless bird becomes horribly vulnerable. The long list of extinct birds has an impressive subsection of flightless species, and they encompass most avian evolutionary possibilities: great auk, Finsch's duck, long-tailed wren, Jamaican ibis, New Zealand owlet-nightjar, Ascension night heron, long-legged bunting.

With flightlessness, there is no limit to size: so a bird can get far bigger than the kori bustard and the mute swan. The ostrich is the biggest surviving bird; the moas of New Zealand were bigger and the elephant bird of Madagascar bigger still. Some birds developed flightlessness despite the presence of predators: ostriches did so in the face of the most fearsome animals in the world. They can cope because they have speed on the ground, size and weaponry: their clawed feet are seriously intimidating. In Australasia, cassowaries have a reputation for attacking dogs and even humans; there have even been fatalities, though these attacks were made by birds that had previously been fed by humans.

The readiness with which birds shed the use of their wings tells us two stories at the same time. The first is the miraculous nature of every bird that retains flight: the high energy demands of that lifestyle clearly put every flying bird on the far edge of evolutionary possibilities. Every sparrow is a small miracle. It is this intuitive sense of the miraculous that has made birds the most studied group of animals on earth: birds, more than any other form of life, are the miracle on your doorstep: the marvel in the back garden. You see a moorhen's labouring, leg-trailing flight as it escapes from the apparent threat of your arrival and realise that it was not a great step for the related takahe or the Inaccessible Island rail to give up flight. Evolution always looks for the easiest option, the most economical way of life. The great extravagance of the albatross or the tiger is, when all the results come in, a piece of precisely calculated parsimony. So it is with flightless birds, which have often evolved on small islands where flight would be a positive disadvantage – so long as the island remained safe.

That was the dodo: standing a metre tall, 3 feet 3 inches, and weighing as much as 18 kg, 40 pounds: perfectly adapted to the gentle life of Mauritius and fearing not a thing in the world. When the environment changes radically, the most perfectly evolved creatures in history will often fail to adapt. As the dinosaurs failed to survive the meteor-strike 65 million years ago, so the dodos failed to survive the arrival of humankind. And so did many birds that had evolved flightlessness. If the humans didn't get them on purpose, the dogs, cats and rats they brought with them would often do the job. On many a flyspeck island, humans had the effect of a meteor and changed everything. Twenty million years of evolution in New Zealand were changed in the course of a millennium – an evolutionary instant – when humans started arriving, a process that accelerated when Europeans joined in.

The biggest bird of prey of them all lived on New Zealand. It could fly all right, fly with immense power and speed – and it evolved to tackle the massive and flightless herbivorous moas which stood 6 feet tall at the highest point of their backs. Female Haast's eagles had a wingspan of up to 10 feet, 3 m (female birds of prey being normally larger than males) which is about the same as the biggest birds of prey still living, like martial eagles and golden eagles. But they were up to 40 per cent heavier, top weight being 15 kg, 33 pounds. That meant that when they built up speed they could hit the enormous moas with massive force. Haast's eagle went extinct around the start of the 15th century because humans had by then killed off most of the moas. The eagles evolved to kill bipedal prey standing 6 foot high: did they also prey on humans? Impossible to believe they didn't give it a go.

But some species of flightless birds, like the kakapo of Zealand, live on, sometimes thanks to massive conservation efforts, sometimes because the conditions they evolved for have not been too radically changed. Stare across the Kalahari, a place where it seems at times you can see forever. That bush in the far, far distance is not a bush: it's an ostrich, lying on the ground and lowering its neck. Ostriches don't hide their heads in the sand in that famously comic way (if non-human life is not portrayed as malevolent, it is generally shown as stupid): but they will drop to the ground to rest up and become extremely hard to see. And watch them going about their business: a male, deep black picked out with white, guarding a dozen chicks fussing around his ankles; standing as tall as 9 feet 2 inches, 2.8 m, weighing 145 kg or 320 pounds and capable of running at 70 kph, 43 mph, and with a kick that can disembowel a lion. Don't call him an evolutionary mistake: at least not to his face.

Valuing oysters

The world is not only weirder than we can imagine; it is unimaginably more finely detailed. So many terms seem perfectly adequate, only to be found wanting after the most cursory look. Monkey? There are about 260 species of monkeys divided into two radically different kinds, those from the old world and those from the new. There are, as we have seen, 21 species of albatross. We think that a term is pretty specific, and then we start to look closer and find that it is far too vague and loose even to be considered a generic term. Oysters are like that. The term is bandied recklessly across the taxonomic groupings of oysters, sweeping up families that are not really all that closely related to each other: pilgrim oysters, saddle oysters and thorny oysters have never had much to do with each other.

The creatures we are most used to calling oysters belong to two quite distinct families. The edible oysters – in itself a loose term, for all oysters are capable of being ingested, but one generally used to refer to the oysters humans most like to eat – are in a quite separate family from the pearl oysters. But that's another confusing term, because practically all shell-wearing molluscs are capable of making pearls.* The pearls we are talking about here are those prized and valued by humans.

The edible oysters come from the various species in the family Ostreidae; the pearl-bearers from the family Pteriidae. The pearl-making process starts not with a grain

---

* The endangered freshwater pearl mussel can live for 250 years.

of sand, as the legend has it, but with the invasion of a parasite. The oyster covers the parasite with nacre or mother-of-pearl, and continues adding layer after layer. The creation of cultivated pearls is triggered by the introduction of a piece of polished mussel shell into the mantle of an oyster. And in only three to six years, you will have your perfect pearl. In the wild, you can go through four tons of oysters and perhaps find three or four perfect pearls. Perfection is of course another human concept: to an oyster, all pearls that relieve the irritation of a parasite are perfect. A pearl is the way an oyster scratches an itch.

Adult oysters generally start off as males producing sperm and then after a couple of years they become egg-producing females; a female can produce as many as 100 million eggs annually, using the scatter-gun reproductive strategy that is operated by many species. (Remember all that it takes for a population to remain stable is for each oyster to produce a single reproducing adult.) Oysters spend two or three weeks as free-swimming larvae before they settle down and become filter-feeders. One oyster can process five litres of water an hour: it follows that a bank of oysters has a powerful effect on the waters around it. Oysters help to maintain water quality, which is a benefit for most living things: it makes sense then, to keep a few of them out there in the sea.*

Oysters often gather in phenomenal numbers, forming reefs, which create habitats for many other species. Oysters are at their most remarkable not on a plate or as producers of jewellery but as an environment.

---

* This crucial maintenance work on our planet was once considered important but incalculable, and therefore irrelevant. It was not a notion capable of slowing down the rate of destruction. A new way of thinking brings together economics and ecology to assess the worth of ecoservices, and attempts to put a serious financial value on them.

Do I contradict myself?

Sometimes I think evolution is just showing off: demonstrating its virtuosity without any sound reason for doing so, like a lead guitarist shredding in a heavy metal band.[*]

It's not just the endlessness of the forms that has this curious mental effect; it's also their deeply contradictory nature. The great edge that birds have – you might say the whole point of being a bird – is that they can fly. Some birds have developed flightlessness, and that works fine until conditions are altered, as we have seen, and that's all very right and proper and understandable. But evolution doesn't stop there. Indeed, evolution doesn't know when to stop: that's almost the definition of the term. And so the forces of evolution have stood all that sort of logic on its head and produced a non-flying bird that's a fish.[†]

We all have a basic idea of what a bird is: a feathery thing that flies. So here is a penguin: a complete contradiction, a smooth thing that swims and can't fly at all. We're back to Walt Whitman; very well, I contradict myself: for nature and evolution are every bit as large even as old Walt, and therefore every bit as capable of self-contradiction. The class of birds has given us hummingbirds that fly backwards like helicopters – you could put a hundred in your hat – and earn a living by flitting from flower to flower in the warmth

---

[*] Old muso's joke: what is the difference between a lead guitarist and a terrorist? You can negotiate with a terrorist...

[†] Of course, evolution has also produced a fish that's a bird: if you have ever witnessed the flight of a flying fish you've seen one of the most joyous contradictions in creation.

of the sun. And penguins. The penguin stands, waddles and swims as living proof of the extraordinary way that nature has of taking things to a logical extreme and then going back in itself and doing something completely different.

But are penguins really so flightless? There are zoos where you can walk through a transparent underwater tunnel and watch the penguins doing their stuff all around you: water is a thicker medium than air, and penguins propel themselves through it with their wings. They are flying all right: it's just that their mobility is *in* rather than over the water: emperor penguins can dive to depths of 565 m, 1,850 feet, and stay submerged for more than 20 minutes. They are as well adapted to the water as seals, and like seals they must come ashore to breed. Eggs tie all birds to land for at least part of their lives: but the great oceans are the places where penguins live most of their lives.

There are 17–20 species of penguins, depending on whose taxonomy you agree with, and they are mostly confined to the southern hemisphere. They are associated in our minds with the Antarctic, but in fact ten species are found in temperate parts of the world and one species, the Galapagos penguin, lives right on the equator. They all follow a very similar body pattern and strategy: the classic penguin (Penguin Classic?) has stout wings that have become flippers, with counter-shaded bodies, dark above, pale below, so that from below they look like sky and from above like sea: a colour scheme designed to foil predators. The cross-ice journeys of king and emperor penguins in Antarctic are famous: the bigger species live in the colder parts of the world. That's a general pattern across the wild world: the bigger you are, the further your vital organs are from the freezing temperatures, and the easier it is to maintain body temperature. It's to do with the ratio of body surface to the stuff inside. Basically, bigger is warmer.[*]

---

[*] A principle known as Bergmann's rule

Penguins are considered endearingly preposterous: comic waddlers across the ice, their upright carriage giving them a droll and faintly human look. They were always a logo waiting to happen. They are creatures that amusingly pop up from the sea like champagne corks when there's a leopard seal in the water. Under the sea they are very different. Their feet, set so far back on their bodies, may make them look funny on land but underwater, that configuration helps them to become flying machines, capable of reaching 12 kph, 7.5 mph, in the water, which is pretty swift in a viscous medium. They can more or less double that when escaping from predators.

The human point of view limits our understanding if we don't use a little imagination. See a penguin from a penguin's point of view – or from a leopard seal's point of view – or for that matter, from the point of view of their preferred prey, krill, squids and fish – and you see an evolutionary triumph of speed, endurance and skill.

One more twist

The rest of the molluscs are gastropods. It's a huge class, second only to the class of insects in terms of species; mind you, that's still a pretty distant second, since practically every animal species is an insect. But gastropods are still impressively diverse, with up to 80,000 known species spread over 202 families. They've also done something that insects have never managed and exploited the riches of the sea.[*] They have also adapted to deserts, and between these extremes, rivers, lakes, marshes, woods, gardens, estuaries, mudflats, oceanic abysses, hydrothermal vents and perhaps the toughest of all habitats, the intertidal zone, where they must cope with twice-daily extremes.

The shelled gastropods are the most famous and most thrillingly various and most wonderfully beautiful. The homes/overcoats they create for themselves out of calcium carbonate are hugely desirable things, and not just for humans seeking collections or cash. I remember sitting on a black sand beach in Colombia with the alarming acid-trip feeling that the entire beach was shifting about in front of my eyes. It appeared to be doing so because it was. Every one of those one-doored gastropod shells was walking about, with a rather random number of legs protruding from beneath in a Dali-esque fashion. The entire beach was alive with moving shells: a quick double-handed scoop would have picked up a couple of dozen with no effort

---

[*] Insects have, however, adapted to the air, something gastropods have never got close to. Flying snails are very rare.

whatsoever. Most were about the size of a thumbnail. These were hermit crabs, crabs who parsimoniously decline to waste their own resources by growing a shell, and instead borrow the discarded shell of a late gastropod. Or possibly a late hermit crab: who knows how many hermit crabs can make use of a single gastropod shell once the original owner has discarded it? The shells were all constructed in the most elegant coiled spiral.

There is a touching sequence in one of the great Modesty Blaise novels,* in which Willie Garvin is emotionally shattered by the ordeal Modesty has gone through. Modesty does all she can to talk him round, and he makes a huge effort to master himself. As a token of his return to his normal state of profound (though never sexual) companionability, he forces himself to speak.

"'Did you know… did you know, Princess… only about one whelk in four million is left-handed?'

Tension drained from her body. She knew she had won.

'Left-handed? No. I didn't know that. I didn't know whelks had any hands, Willie.'

'Not 'ands, Princess. Their shells are twisted right-'anded. You could work all your life with whelks, and the chances are you'd never see one with a left-'anded shell.'"

And then they get off together and kick hell out of the bad guys.†

The shelled legions of the gastropods are things of wonder: ribbon bullia, rugged vitularia, Australian trumpet, zebra volute, distorted triumphis, bubble marginella, large perverse turrid, Japanese wonder shell. But there are many

---

* *Sabretooth* by Peter O'Donnell, first published 1966.
† Fascinatingly, this is not the only shellfish reference in the major thrillers of the 1960s. James Bond finds Honeychile Rider in *Dr No* after she has been gathering shellfish; the chapter is called "The Elegant Venus". Honey explains, with fine appropriateness, that she has been gathering specimens of *Venus elegans*. This is a real shellfish, though the name is now obsolete; it is a bivalve, rather than a gastropod. In the film Ursula Andress, as Honey, makes her famous entrance carrying two conches; these of course *are* gastropods.

gastropods that don't have shells, or have internal shells, or very small shells. They manage to survive in environments that can't provide calcium carbonate to make shells, and so they simply do without.

We have already met the slugs in their sexy extravagance. There is a quite different group called sea slugs, which can be beautiful in their own right, coloured vibrantly and gliding through the water breathing through feathery, clearly visible gills. They are called the nudibranchs, because they have naked gills. They too come in some considerable diversity: the Spanish dancer in deep crimson, the elegant sea slug in pink and yellow, Anna's sea slug in purple and blue with white and yellow, as lovely a slug as anybody is capable of imagining.

There is also a mystifying group called sea hares, which are related to sea slugs but have long sensory head-stalks, which you could say look a bit like ears, so they make a kind of acid-stoned maritime hare. There is another distinct group called the swimming sea slugs, and they include a species called the common sea angel.

It's not a shell that gastropods have in common, then: it's the torsion. In the course of their development, each individual gastropod twists along the head-to-tail axis. By doing so it loses some aspects of its bilateral symmetry, which is a puzzler, because though technically a bilateran, it is not actually bilaterally symmetrical; the wild world is always capable of muddling any certainty, making yet one more variation, one more contradiction, one more twist in the plot. The torsion has nothing to do with Willie's twisting of the shell: all gastropods do it, shelled or not.

Hijoputido

It was the kind of sinister weather you can get on the coast of East Anglia: dark sky, a marginally darker sea and a wind that cut like a razor. There was a feeling that the sky was capable of any enormity: white-out, blizzard, face-stripping sleet, concussive hail, rain that might penetrate any number of layers. My fingers were clamped round the binoculars in a death-grip and I wondered how long the skin on my face would stay put. This was sinister weather in a sinister place: Shingle Street, which has strange legends of a failed German invasion during World War Two and of a pub destroyed by a bomb developed at Porton Down.

And in an instant there were 60 birds flying around me. Extremely ordinary little birds, bounding rather in their flight, with odd flashes of white. They seemed quite a mixed bag, a species that was sloppily carrying on without a proper uniform. A non-birder would have thought nothing of them: wondering only what these sparrows were doing on the beach. And further north, much further north, the same birds happily fill the house sparrow niche around Inuit settlements – so for them, the weather was balmy: certainly conditions in which they were relaxed and confident. They were snow buntings, the world's most northerly breeding bird, perfectly capable of raising a family up to and beyond the 80 degree parallel. And they are passerines: but then most birds are passerines.

As you look through the Animal Kingdom, every now and then you come across a basic design that works incredibly well. Well, all designs of living species have to be said to work, or the creature wouldn't be alive: what I am getting at here is that some designs are capable of immense variation.* We looked at gastropods in the previous chapter: we will look at the same principle again, and again and again when we come to insects. Among mammals, rodents provide the biggest variation: but there are more species in the order of passerines than there are in the entire class of mammals: getting on for 6,000. About 60 per cent of all bird species are passerines.

They tend to be roughly the same shape, a sort of bird-shaped bird, if you see what I mean. A few of them, like the whydahs, carry spectacular decorations, but stripped down, the body plan is largely similar to that of their fellow passerines. The USP is in the legs. Passerines all have feet with three toes in the front and one behind – standard bird layout, really – and they all have a muscular mechanism that locks the foot closed. This means that they can perch: but better than that, it means that they can sleep perched without falling off. From my place in the winter I often see large numbers of jackdaws and rooks coming in to roost, wheeling, jacking and cawing, as they take their places in the summits of half a dozen trees. Here they will sleep the night. And not fall off.

In Britain the passerines range in size from the raven to the goldcrest: globally the range of size is much the same, the prizes for the extremes going to the thick-billed raven at 1.5 kg, 3.3 pounds, and the short-tailed pygmy tyrant at 4.2 g, 0.15 ounces. And they come in their thousands. The small brown bird that you can't identify is always a passerine: British birders call them LBJs, or Little

---

* This is not an unvarying pattern, of course. There are some species without close relatives, some creatures that are more or less a one-off. There are some mammals and some birds in which one species constitutes not just the genus but the entire family: the family is, then, monotypic. These misfits include some of the weirdest creatures on the planet: platypus, koala, aardvark, pangolin and the Ganges and Indus river dolphin; and in birds, the hammerkop, oilbird, shoebill and wallcreeper.

Brown Jobs; Spanish birders, more imaginatively, call them *hijoputido*, a bird from the order of little bastard passerines.[*]

Most of the birds on your bird-table are passerines: robins, blackbirds, thrushes, greenfinches, goldfinches, linnets, nuthatches, sparrows, dunnocks, starlings, great tits and blue tits. The only exceptions will be the pigeons and doves, and if you're lucky a great spotted woodpecker. Just about everything is a passerine. The order includes birds as exquisite as the firecrest, moving like a clockwork toy with its head on fire through the pine needles, and species as commonplace as the sparrow.[†] There are passerines that people cross the world for, like the birds of paradise; there are passerines that even the most obsessive birders wouldn't cross the street for. I have sat in a rainforest in Sri Lanka while the astonishingly confiding Sri Lankan blue magpie, one of the island's many endemic[‡] species, got on with their business all around me;[§] I have walked for miles along Cornish cliffs with the sea on one side and the skylarks and meadow pipits on the other, forever with me.

I have stood in a Suffolk marsh and seen one of the great sights in British birding, perhaps in world birding, when a starling flock performs its nightly miracle. The magnificent aerial ballet – a murmuration – seems to be a celebration of starlingkind: an expression of togetherness, a way of

---

[*] *Hijo de puta*, son of a whore, is a routine Hispanic insult. David Beckham was sent off during his Real Madrid days for trying it out on a linesman.

[†] House sparrows have declined, most notably in London, from super-abundance to relative obscurity. There are many suggested causes for this, but no one has nailed it. Probably it comes down to a complex suite of causes. This is worrying because the more complex a situation, the harder it is to reverse it.

[‡] An endemic species is a species confined to a single place; it is most frequently used (especially in birding circles) to refer to an animal that lives only in one country. Zambia's only endemic bird is Chaplin's barbet, which has been patriotically renamed Zambian barbet.

[§] Many Sri Lankan birds are astonishingly bold and will perform for visiting humans without self-consciousness. This gratifying trait has been put down to the country's Buddhist history. The suggestion is that the traditional respect for non-human forms of life has given birds an unusual confidence in human beings, and with it a reduction in the traditional flight distance.

strengthening and revelling in the immense sense of contentment and safety that comes from the great rich numbers of the flock. And I remember my first morning in Africa, stepping from my hut by the Kafue River and finding not sparrows on my doorstep but birds of the most improbable shade of powder blue: blue waxbills, welcoming me to a world of altered possibilities. I was never quite the same after the waxbills.

All these are passerines: and passerines go on and on, many of them discreet and secretive and hard to identify in their littleness and their brownness, others taking a more extravagant line. Passerines are everywhere and always with us, and as such, they set the tone for the places we live in and the places we visit and the places we pass through. Anywhere we can get away from concrete, passerines define our world, and even among the buildings, there are passerines. Pied wagtails in Britain have a great affinity with car parks and flat roofs and will dance* across them in search of insects. But any open space, with a bit of earth or grass or shrubs or trees, will take its character and its meaning from the passerines that make use of it. More than any other group in the entire Animal Kingdom, more than any other group in the entire class of birds, passerines define the place and the time. There is a sense in which passerines define ourselves.

* For an Italian, every wagtail is a *ballerina*.

Creeping like snail

The contemplation of extremes is always helpful. The best way to do the contemplating is by being there, sharing the same space, clapping eyes on, sometimes touching, sometimes even picking up and holding. It's all very well knowing that the blue whale has a tongue the size of an elephant: it's quite another to see a blue whale for yourself. And in the same way, it's amusing to read about the bumblebee bat – sometimes called Kitti's hog-nosed bat – that's found in Thailand and Burma, but when one of those absurdly tiny little things flies over your head it adds another layer to your understanding. You need that sense of wonder if you want to reach for the secrets of the universe, and the gosh response of the child – and for that matter, the fuck-me response of the non-cynical adult – is the best way to nurture the process. And very few things can do this quite as simply and straightforwardly as size. I have, for example, sat in a canoe looking *up* at drinking elephants, three massively ivoried bulls who would have been impressive at any angle. But from a few yards away and in a position that started lower than the feet of the elephants, I was impressed as never before by the sheer elephant-ness of elephants: the ah-ness of elephants, if you prefer: the trunks as they are, the tusks as they are.

So here's yet another of those fuck-me moments that a walk across the African savannah will hand you with such reckless generosity. It concerns something much smaller than an elephant, but it produces the same head-wagging sense of wonder and privilege: the same not-quite-disbelieving revelation of the liveliness of life. Stoop for a moment and pick up a shell. The gastropod body has long left, but what you hold in your hand is something you will always remember.

It's the shell of one of the great African land snails: a group of species that to most of us look impossibly huge. To give the dimensions doesn't really do the experience justice: say, 8 inches or 20 cm long, and up to 3 inches or 8 cm high. There have been some even bigger individuals: a length of 30 cm or a foot, and a diameter of 6 inches or 15 cm. But it's holding it in your hand that matters: the feeling that this can't be right, that snails simply aren't that big. Imagine a song thrush trying to hammer this baby against a stone: the song thrush could practically creep inside the (right-handed) spiral of the shell. It's an experience that says something about the brilliance, and the limits of the gastropod body plan. Oh brave old world that has such gastropods in it!

The land-dwelling shell-wearing gastropods are familiar to all of us: we give the whole lot of them the name of snails and believe that that is the end of the matter. But an encounter with one of these African monsters destroys this complacency. We live in a world of wonders whether we like it or not: and no matter how we view it there are very many more kinds of everything than our minds can cope with.

Stephen Jay Gould, the great writer on evolution and related topics mentioned elsewhere in these pages, was especially devoted to snails. Like Darwin with his barnacles and his earthworms, Gould believed that big ideas come from minute studies. He had a special fascination for the West Indian snails of the genus *Cerion*, the molluscs the Jamaican shell expert Honeychile Rider forgot. Here's Gould:

"*Cerion* is the land snail of maximal diversity in form throughout the entire world. There are 600 described species

9. GIANT AFRICAN LAND SNAIL

of this single genus. In fact, they are not really species, they all interbreed, but the names exist to express a real phenomenon which is the incredible morphological diversity. Some are shaped like golf balls, some are shaped like pencils*... Now my main subject is the evolution of form, and the problem of how it is that you can get this diversity amid so little genetic difference, so far as you can tell, is a very interesting one. And if we could solve this we would learn something general about the evolution of form."

The key, then, to one of the most important questions about the development of life on earth is to be found in these snails. These snails hold the meaning of life... in common with every other species I have mentioned in this book, in common with all life on earth.

---

* Gould suggested that if only Columbus had picked up a *Cerion* shell and kept it safe, we would know exactly where he first landed in the (brave) new world.

Wild thing

An area of wet scrub in Norfolk: reeds, dense vegetation, behind it some taller shrubs. And from the scrub a song. Was it reed warbler or sedge warbler? I paused to listen, for this was a decent singer, and either species is worth a listen. It was mid-June, late in the year for such a tremendous burst of sound. The conundrum of the singer's identity began to intrigue me: the pair should be pretty straightforward to tell apart, but this bird seemed to be singing like both of them. And there was more to the song than there should be. It was not as rhythmic as the reed warbler ought to be, and rather greater in range and imagination than the sedge usually is. And were there other birds in there with him? And at least one, no, two singers I didn't recognise at all. Or did I? There was something familiar going on, but increasingly the bird – surely there was only a single singer – had me baffled.

It was my companion, Carl Chapman, who runs Wildlife Tours and Education from Norfolk, who cracked it. Marsh warbler. A bird that comes to England to breed in tiny numbers – 20 to 40 pairs a year has been suggested – and to do a spot of singing. It's a classic LBJ, Little Brown Job, member of the *hijoputido* group. But when it sings it becomes one of the most glorious, extravagant and over-the-top birds on the planet. It can sing for an hour or more without a break, and in that time create a song of such wit and invention that it leaves you staggered. It's most famous for incorporating into its repertoires the songs of other species, and twisting and turning them to its own purpose. One famous piece of research on a small population counted a total of 212 different species imitated, each male having an average repertoire of 76. You need quite an ear to pick them all out, for they are jumbled together to conform to the bird's individual notions of music. The whole business is complicated by the fact that the bird spends its winters in Africa and picks up at least half of its repertoire down there. The 12 European birds that it is most commonly heard to imitate are blackbird, house sparrow, tree sparrow, whitethroat, swallow, blue tit, linnet, skylark, starling,* stonechat, great tit and magpie. But these can be hard to pick out among all the Africans, many birds I am deeply familiar with. One observer heard a chaffinch's characteristic *spink*[†] call melding into the call of a puffback from Africa, and a blackbird song woven into that of white helmet-shrike, another African.

The marsh warbler is one of the great singers of the world. The entire membership of the group of warblers are pretty much LBJs, sometimes with thrilling bits of olive-green thrown in. They are inconspicuous and to a considerable extent indistinguishable. They sing their colours instead: they sing their identity as a species, and the more complex singers among them sing their own individuality as well. Some birds like the chaffinch – not a warbler – sing the same phrase endlessly and seem never to tire of it; others like the marsh warbler, the nightingale, the skylark and the song thrush are inventive and creative and individual. So much so that they seem to sing for something beyond the hardcore Darwinian reasons of procreation. Biologically, yes, a bird sings to establish and defend a territory, to find and protect a mate and then a brood. But I suspect that they sing also because it is in their nature to sing. They sing because they must: because the music in them needs to

---

* Which is itself a famous mimic.
† The call, usually transliterated as *spink*, can also be written as *finch*. As such it gives the name to the entire family to which it belongs.

escape. Marsh warblers love to sing and three or four males will on occasions form a group and sing together, softly and more economically than when they are protecting a territory. It seems they are singing for the pleasure of it, for there is no apparent biological function to these glee parties. They just enjoy singing, singing together, sharing material and making music. This is nothing less than a jam session: and I hope you like jammin' too. The best singers among a population of songbirds, the most inventive of the repertoire singers, tend to get the best territories and the best mates: but surely when a bird sings, it sings the song of itself, rather than merely sending out a biological signal. Birdsong, like the peacock's tail, is more beautiful than it needs to be in terms of mere biological function. The great avian singers are surely creative artists.

Most birdsong comes from the passerines, and of these, most come from the oscines, a group that is often loosely referred to as songbirds, for all that birds which are not songbirds also sing. And this bounty of song has provided the soundtrack of planet earth. I have suggested elsewhere[*] that birds gave us more than half our music. Rhythm is our birthright as mammals: we humans spend the first nine months of our existence in thrall to the 4/4 heartbeat rhythm of our mothers. But for melody, we turned to the birds, and the first human instruments were bird flutes, frequently made from the hollow bones of birds to add to their magic.

There is a particular intimacy about the relationship between birds and humans. Birds are daytime flaunters while mammals are so frequently nocturnal skulkers. Most mammals operate by smell, which is a sense we are incapable of understanding as a dog does. But birds are mostly creatures of light[†] and colour and sound, and we have lived

---

[*] *Birdwatching with Your Eyes Closed*, published by Short Books, 2010.
[†] "Ineluctable modality of the visible": start of the third chapter of *Ulysses*.

alongside them and we have looked on them as creatures that enhance our lives from the dawn of human existence on the African savannahs – where our ancestors doubtless imitated the songs of the birds and joined them together and improvised on them and jammed with them as a marsh warbler does. We humans are mammals, but in many ways we have a closer relationship with a quite different order in the same phylum. The song that shares its title and its opening line with this chapter heading* continues: "You make my heart sing!" Birds taught humans to sing, to express our joys in song. Birds provide many ecoservices for humankind, but perhaps the most important of them all is right here. They sing: they make our hearts sing.

---

* "Wild Thing", written by Chip Taylor in 1965, was covered by The Troggs the following year. It was performed, perhaps definitively, at Monterrey Pop Festival in 1967 by Jimi Hendrix.

On our last legs

Here, then, is the last phylum to be introduced in this book. I've devoted an awful lot of the words in the vertebrate cycle of this book to us fellow mammals, but that's because I'm a mammalian chauvinist pig. I must devote every bit of the rest of the invertebrate cycle to the arthropods, because I have no option. Get ready, then, for legs. Lots and lots of legs. Arthropods go in for legs in sometimes reckless numbers, constantly reminding us how little we have in common with them. Arthropods include crustaceans, that is to say lobsters and crabs and the woodlice in your back garden. They also include a weird little group called sea spiders. They include the arachnids, which are far more than just spiders. Then there are the curious horseshoe crabs. And – oh yes – there are insects. Endless, endless, endless forms of insects, some of them most beautiful, others rather less so, at least to my eyes, and no doubt to many others', though lovely enough to their conspecifics.

Arthropods are fantastically variable. Never mind this being the Age of Mammals, or the Age of Humans: as said before, this is the Age of Arthropods, and it has been so from one geological era to the next. The earliest arthropod fossils are – though this is naturally disputed – pre-Cambrian, dating back to 555 millions years ago. Arthropods surpass every other phylum in the Animal Kingdom as greatest does least. Of all the animal species described, 80 per cent are arthropods. They comprise more than 2,500 families, and about 1.2 million species. Those, at least, are the ones already known: you can make your own estimate about how many remain to be described, bearing in mind that a top-quality field naturalist can find a new species with every saunter in the rainforest, on more or less every tree in the rainforest. Some go as high as ten million; others will bid still higher. It has been said that we don't even know what order of magnitiude* to think in. That's the concept that really hits home, I think; the concept that reveals so many of the obscure truths about life on earth: that there are so many different kinds of animals that we don't know whether to think in thousands or millions or billions.

What is it that unites the giant crabs, the horrible hairy spiders of living nightmare, the city-building termites, the incomparable butterflies and the plaguey flies? First it's the tough outside of them. All arthropods wear a cuticle made of chitin and protein. This makes up an exoskeleton: the arthropod's equivalent of Le Corbusier's machine for living in. The second thing is the segmented body. The number of segments varies, but the idea of dividing the body up into individual chunks is common to all arthropods. And after that it's the jointed appendages: the sticky-out things that are also armoured in chitin, but can move and flex because of the ingenious way they are hinged. Appendages can include antennae, mouthparts, sexual bits and – yes, that's right, legs.

Let me tell you about an experience – an epiphany – that involved 376 legs, 378 if you include my own, even if I was pretty legless at the heart of the story. Of all my African boasts – and as you know to your cost, they are many – this is the one that, I think, shows the greatest courage: the sort of courage you can always summon up when you have absolutely no choice in the matter. You give it the name "courage" just to encourage yourself. I was staying with friends in Zambia and it had been a good day. I had ridden a horse across their ranch, which was enormous and full

---

* A different order of magnitude comes with each additional nought.

of antelopes and mature trees. The meal at the house had been good, the after-meal was long and whiskified and full of talk and laughter and good tales. I was staying about a mile down the road in one of their guest huts; I was driven there, it not being safe to walk about. I said my goodnight, opened the door and turned on the light.

Now this was a well set-up hut, with insect gauze over the windows and so forth. No need, then, for a mosquito net. And on the walls and ceiling there were a few spiders. I shall be frank: I don't care for spiders. I don't run screaming from them, but I'm not happy about them. I can pick up a moth in my cupped hands and feel its tickling struggles as I transport it to safety: I couldn't think of attempting any such manoeuvre with a spider. And these were not the modest little things you get in a house or a garden in England. They were serious. They were from the group of wolf spiders: not web-builders but spiders that lie in wait and jump on stuff. They ranged in size from a 3-inch to a 6-inch leg-span, 8–15 cm, with the emphasis on the latter. Not that I put a ruler on them. But I did go as far as to lie on the bed and count them: eight legs to each one (even if quite a lot of the legs seemed to be worth double), and that made, as I say, 376 legs in all. Which divided neatly enough into 47 spiders. I think this counted as a fuck-me moment. Perhaps also a fuck-off moment. Certainly it was some kind of epiphany. So I had a choice. Or to put it another way, no choice. I could in theory face possible injury and death by walking back across the bush, and certain ignominy if I survived, or I could deal with my own not-quite-gibbering terror.

I didn't need a drink. There was plenty of whisky in the bloodstream already. But just for the look of the thing, I poured myself a slug from the duty-free, mixed it with a dash of water and sat there drinking and staring and counting and meditating on the phenomenon of legs. And my conclusion was that arthropods are the world masters of legs: that legs are the way in which they deal with the problems of the world: legs in multiplicity. Two or four is no good at all for them. An insect insists on six, a spider on eight, 750 have been counted on one of the giant millipede species.

I finished my tautologous whisky. Extinguished the light. That was the brave bit. Shamelessly pulled the sheet over my head: I knew they wouldn't *actually* drop on my face like little, hairy, wriggling, giggling and jiggling bombs in the night, but I didn't want to give them the option. And I slept the sleep of the just and the courageous and the pissed until morning. A sore head, but greatly enriched all the same. It's not every day that you get a chance to contemplate legs in such profusion: not every day that the wild world forces you to spend so much time thinking about the world of invertebrates. And their teeming, scuttling, multiform, multipurpose brilliance.*

---

* There is a species of wolf spider that lives in the volcanic tubes of Hawaii, where it is in some danger because the tubes are drying out. Water extraction for farming is the problem. They come from a group known as big-eyed wolf spiders, for the obvious reason, but these cave spiders, though clearly related, are in fact blind. So gloriously, they are known as the no-eyed big-eyed wolf spider.

Blood-chilling

There was a moment when I stood precisely on the line that divides mammals from reptiles. Us warm-blooders from them cold-blooders. The hot-blood–cold-blood cusp. It certainly chilled my blood. I was on the banks of the Grameti River in Tanzania, one of the rivers that cross the Serengeti. It was an idyllic scene: like one of those Constable landscapes in which nothing can ever go wrong. A line of trees followed the banks of the river, the canopy closing over the middle of the narrow stream so that the river ran through a green tent. It was pretty narrow at this point, no more than 20 yards. From the canopy, two of the sweetest voices in Africa, soft and subtle: the orange-breasted bush-shrike, singing charming improvisations around the first few notes of Beethoven's Fifth, and the black-naped oriole with his liquid whistle. It was a landscape, a soundscape of perfect peace, annihilating all that's made to a green thought in a green shade.

I left the vehicle and walked – almost – to the water's edge. It looked the perfect place to take off your clothes and walk into the water in the company of a couple of girls from a Gauguin painting. Until you looked at the water. Not logs, no. Crocs. Enormous crocodiles. Three or four of them, half a dozen, no, more, look upstream, look downstream. A lot of very big crocs in a very small river. Now if you are used to looking at wildlife you start to understand something about what's fitting and what's not, and this was not fitting. This gathering of enormous crocodiles was too much for a river so small and so charming. There was not enough food here for a dozen or more full-sized 16-foot, 5-metre, crocodiles. There had to be something else. And it was me.

Or us. As I stood by the river, I could see very clearly that the crocs were looking at me. Looking at me with great interest. They didn't hunt for their food in the water: what's the odd fish to a beast that size? They expected food to come to them from the land, and they expected its blood to be warm.

I had spent the last few days with wildebeest, celebrating the extraordinary business of their migration. And then I travelled onwards, getting effectively two weeks ahead of the migration front, and that brought me as far as the Grameti River. In a fortnight's time, then, the wildebeest would be crossing. There they would meet the crocodiles, the crocodiles who hadn't eaten since the last time the wildebeest came through, the crocodiles who hadn't eaten for a year. And they were very bloody hungry. A year's wait between meals, but then it was bonanza time, a fortnight or so with an endless conveyer belt of food. Soon the buffet would arrive, each dish calling to another with the great frog-bleat of the wildebeest: *Newp! Newp!* No fish for these crocs. Mammals. Us lot. The crocs possess a body that can operate on slo-mo, sitting out days and days of quiet and starvation, waiting, always waiting, waiting without any concept of patience, still less of impatience. Boredom? What's that to a reptile? What's a year between meals, what's 50 weeks of trying to catch the waiter's eye if the meal he eventually brings you is so exactly what you ordered?

We warm-blooders, we mammals and birds, are high maintenance, and we have a high concept of busyness and a low tolerance of boredom. We have a high metabolism: we need to be busy in order to stay alive and our way of staying alive makes us constantly busy. We can't afford to wait: if we wait, we die. Some shrews must eat almost all the time they are awake. Big carnivores like lions spend a fair amount of time dozing and cosying up to each other but if they don't eat every couple of days or so they start to starve. Our mammalian, our warm-blooded concepts of time and life are radically different from those of the crocodile. Mammals and crocs live in the same world in three dimensions, but when it comes to the fourth dimension of time they are aliens. As I stood on the banks of the Grameti, I stood on what divides us. Step a little closer, they said, and we will move in a great deal less than a second from stillness to the most violent action. The fact that we haven't moved much for 11 months and more doesn't mean we can't do so at a nanosecond's notice. Try us. Step ten paces forward and pause for an instant of time while the water laps at your toes. Do you think you're still out of range? Try us and see.

And then the sweet sibilant sounds of this watery glade were joined by a strange sort of gargling growl. Yes, mammal, we do mean you. You personally. I returned to the vehicle panting as if I'd just run the 100 yards.

Those are reptiles, then: vertebrates like us, but quite alien. Or are they? The concept of cold-bloodedness is deeply strange to us. How can we begin to imagine a life in which most of the body heat is generated from the outside rather than the inside? But reptiles are linked with birds and mammals by the fact that we are all amniotes, unlike other vertebrates. We have in common an amnion, a double membrane which permits reptile eggs to be laid on land rather than in water. That holds good for birds as well, and also for the egg-laying mammals, platypus and echidna. And confusingly, it is also true of mammals like humans. The foetus, carried within the female, is protected by several membranes, including the amniotic sac, which contains the foetus itself. Despite our radical differences, we are also one with the reptiles.

Modern taxonomists haven't entirely settled the way

the traditional class of reptiles should be ordered. The fact is that some reptiles – the crocodilians, for example – are more closely related to birds than they are to other reptiles. Some taxonomists have abandoned the concept of reptiles altogether, or redefined it to include birds. And if you look at some baby birds, those that don't hatch out good-to-go,* you will see that they look unmistakably reptilian. A clutch of bird-of-prey chicks looks like a handful of little dinosaurs. Reptiles have been defined as "non-mammalian non-avian amniotes". Crocodiles, us – what's the difference?

But on an intuitive level the difference is very clear, just as the similarity is very clear when you meet one of the great whales and hear the great breath. Reptiles may be like us in some ways – number of limbs, backbone, amnion – and we certainly have an ancestor in common. But they don't live in the same way, they don't see life in the same way, they don't think the same way about food and time and action and rest. A croc can afford to wait. I've seen 100 at a time in dry-season gatherings in the Luangwa River, in such deep pools as are left, outlasting the oppressive weather until the freedom of the rains comes: a sight straight from the Jurassic, for crocodilians have changed little since the time of the dinosaurs. I've seen them strike: or rather, I haven't seen them strike. It seems that a croc has two speeds, slow and warp. Stand close to a croc and you are aware of a deep truth: that crocs, that reptiles, are a different class.

---

* Newly hatched birds come in two kinds, precocial and altricial. The first are capable of moving around soon after they are hatched, the latter kind are not, and are pretty much helpless. Precocial chicks look like fluffy darlings; altricial chicks look like little (reptilian) monsters.

A suit of armour

My grandfather took me several times to Birmingham Museum. I was aware that it would be tactless to compare anything in Birmingham unfavourably with anything in London, so I didn't say that the natural history collection was a bit thin compared with the great treasure-house in South Kensington where I had my early education. I can remember only two of the exhibits with any clarity. The first was an excellent Triceratops skull, the second a Japanese spider crab. This crab was displayed vertically: pinky-orange and with a leg-span of about 12 feet, getting on for 4 m. It was leggy to a disturbing degree. I couldn't help throwing sneaky glances at it after we had passed it, in case it should come scuttling and rattling after us. Not that it could have done so even had it been alive: it needs the support of water to make those vast limbs and those endless pincers operate effectively. There was an unnatural look to it for that reason: it looked like a creature of diseased fantasy, for all that it was quite obviously real.

This is a crustacean, and the largest of all the arthropods. Crustaceans include crabs, lobsters, crayfish, krill and barnacles. Western civilisation regards many crustaceans as supremely edible but turns up its nose at their fellow-Arthropods. Insects,* for example. Insects have nothing to

---

* But see the 1895 pamphlet by Vincent M Holt *Why Not Eat Insects?* Holt writes: "What a pleasant change for the labourer's unvarying meal of bread, lard and bacon or bread and lard without bacon, would be a good dish of fried cockchafers or grasshoppers?" More recently, the Dutch scientist Marcel Dicke has recommended the consumption of insects for reasons of nutrition and ecological sense. "Locusts are nice cooked with garlic and herbs." Amusing,

do with the sea, but the oceans are full of arthropods: there are crustaceans everywhere in the oceans. There are even a few land-dwelling crustaceans, like woodlice and some species of crab, and a few crustaceans are parasitic. Around 67,000 species have been described, but there are certainly many more awaiting discovery. Again, even the order of magnitude is in doubt: there may be ten times or 100 times more species.

Their stout suit of armour allows these somewhat ponderous and often slow-moving creatures to survive and prosper. They moult this exoskeleton as they grow. They are defined by two-part limbs and by the nature of their larvae. Krill[*] form a group more remarkable for their numbers than practically any other on earth. The stats defy belief: it has been estimated that the biomass[†] of Antarctic krill totals 500 million tonnes: twice that of all humans on the planet. They can swarm at densities as great at 60,000 individuals in a square metre. These tiny creatures are food for other animals, many of them much larger: fish, squid, penguins, seals and the baleen whales. That the blue whale can survive by eating things that look like tiny shrimps is one of the head-spinning paradoxes of the wild world, and it baffles me as much as it did when I first learned such things in the Natural History Museum in London. Not that krill see themselves as a swimming larder: they have complex and sophisticated ways of evading predation, which involves both swarming and scattering techniques. Some krill have been observed performing an instant moult, leaving their discarded exoskeleton behind as a decoy. They filter-feed, mainly on phytoplankton: that is to say, planktonic species

---

I know. Is it more amusing to smash up the rainforest to supply the world with beef?

[*] The term krill is a little imprecise. It has sometimes been used to cover all kinds of planktonic animals, including small fish and fish-fry. Here it is used for the species of crustaceans that are found in the seas in such mind-spinning numbers.

[†] The total dry mass of an animal or plant population – *Larousse Dictionary of Science and Technology*.

that operate by means of photosynthesis, rather than eating stuff – plants, in other words, and therefore different from animal plankton (zooplankton to be more technical). Ecologically, krill convert an unreachable and inedible resource into an edible form, which is themselves.

A further paradox of the wild world is that the prey population controls the predator population: the apex predators are always the most vulnerable animals in the food chain. Without krill there are no whales: so these little specks of marine life are controlling the largest animal that ever existed. Krill are harvested by humans for aquaculture and aquariums and as bait; they are eaten by humans in Russia and Japan. Climate change is having complex effects on the world's oceans and it is possible that krill are vulnerable to this. Krill more or less hold the oceans together: their vulnerability is a looming problem for humans as well as whales.

Snakes, unclad humans
and a garden

I would sooner face a poisonous snake in my living room than a serious spider. I would do so naked and suffer fewer terrors than a harmless spider would give me. In fact, I have done so, and on more than one occasion; I used to live on Lamma Island, one of Hong Kong's outlying islands, and snakes would occasionally come visiting. On two occasions these were bamboo vipers: a sumptuous green picked out with scarlet around the tail and a fine yellow underneath. They look exactly as poisonous snakes should. They are unlikely to kill you unless you are a child or have a weak heart, but all the same, anyone who got bitten was supposed to get helicoptered off the island. So they are serious snakes. I don't claim any particular bravery in dealing with them: it's all in the way these things take you. I was, as it were, *rationally* afraid of the snake, so I was careful. But big spiders, no matter how harmless, all give me the willies. I have known people express a still deeper terror of snakes: often people who have never clapped eyes on one. Snakes don't kill as many people as you might think: most people who get bitten by a snake have been handing them, so there's a hint.

There's more complex mythology connected with snakes than with any other group of creatures on the planet. In western culture snakes are associated with the unapologetically evil, from the Book of Genesis to Harry Potter. Phallic, poisonous, legless, undulating, slimy, sinister, cold-blooded, unblinking: they stand for the wild world at its most unhuman and therefore most threatening. As a logical extension of this, snakes have come to represent all the evil in the world, everything that is dangerous and disturbing.

Which is a tough load for any creature to bear. It's the poison thing that does it, I suppose. Actually the correct word is venom: poison is ingested or inhaled, venom is injected. Intuitive human taxonomy places snakes in two classes: venomous and non-venomous. There are around 3,000 species of snakes, and less than a quarter of them are venomous; of these, only about a third are capable of killing a human. However, it has to be admitted that the idea of 250 species all coming to get you at the same time is as compelling as it is irrational. It was a situation of high alarm and low farce when I found a snake in the living room: naked – they always seemed to be discovered first thing in the morning – I herded it into the middle with a broom. My then girlfriend, now my wife, was in the same state – there's scriptural precedent for encountering snakes in this fashion – but without a broom. She was helpfully leafing through *A Colour Guide to Hong Kong Animals*, the very copy of which I consulted a few moments ago (she gave it to me) to check the colouration of the bamboo viper. "I think it's this one, which means it's not poisonous."

"I think isn't really good enough, Cind..."

Anyway, it turned out to be harmless and I swept it into the tiny front garden to sort itself out.

But snakes are not catalogued this way by zoologists, who divide them into 18 families. Venomous species crop up in three of these families, the elapids, which include cobras, kraits, mambas and sea snakes; the viperids, which contains vipers, rattlesnakes, cottonmouths and bushmasters; and colubrids, which include boomslangs. Most colubrids are non-venomous. I remember hearing a crashing sound on a drive through the bush at night in Malawi. We eventually found that a black mamba – it must have been at least 10 foot, 3 m long, though I didn't get out and measure it – had fallen from the tree it had been creeping about in. Its truly colossal weight had been too much for the thin branches it had trusted itself to, and down it came. It lay out draped across a series of bushes like tinsel around a Christmas tree. It takes a lot for a snake to lose its dignity, but this one looked ridiculous. Which didn't mean it couldn't still kill you.

Venom is for disabling prey, not for menacing humans. Most snakes kill by other means: some by constricting, others by swallowing live prey,* all snakes being carnivorous. All snakes have many joints in their skulls, which means they can swallow prey larger than their heads, a particularly unnerving talent. They range in size from the thread snake, which is 4 inches, 10 cm long, to the reticulated python, which can get close to 10 m, 33 feet. There is a fossil snake, splendidly named *Titanoboa cerrejonensis*, that reached 15 m, almost 50 feet. Snakes probably evolved from lizards, but aren't to be confused with legless lizards, like the slow worm, the lovely copper-coloured reptiles that can be found in this country if you get lucky and keep your eyes open. Legless lizards have eyelids and external ears: the snakes' lack of such things gives them their singularly snake-like look; their eyes are covered by transparent protective scales. As arthropods have their being in legs, a snake has its own in length. Everything about a snake is about being long: so much so that many species have only one functioning lung, and paired organs, like the kidneys, are not placed side by side, like ours, but fore and aft to economise on width.

Their uncanny and unnerving reputation is enhanced by the senses they possess: an ability to smell with their tongues: more or less tasting the air, which is why they

---

* Egg snakes kill their prey at a slightly earlier stage of development, by swallowing an egg whole and then crushing it.

go in for the tongue-flickering. They repeatedly touch the tongue's fork to receptors in the mouth to process information about the environment. Some species are equipped with infra-red sensors which allow them to home in on the heat produced by a warm-blooded vertebrate. They are all acutely sensitive to vibration. They have more than one way of moving, though not all snakes can do all of them. They can operate by lateral undulation on land, and a different version of the same technique in water. They can side-wind when there is nothing to push again (a great favourite of natural history documentaries, this) and they use a concertina motion when there is not enough room to side-wind. They can also move just by willing themselves forward – well, that's what it looks like. Even from up very close, you can't really see how the trick's done: they just flow along. It's called rectilinear motion: they lift and pull forward the scales of the belly and drag themselves along in a slow, effortless and wonderfully spooky way. Tree snakes use a complex combination of these techniques to climb. And some snakes can take to the air at will: the gliding snakes of Southeast Asia can spread their ribs to create an effective aerofoil surface. They can cover several hundred feet in the air and can even change direction while doing so.

Oh, and they're not slimy. They are creatures of singular beauty that for some reason have given humans the willies across countless millennia.

Beloved barnacles

Brilliance is not enough. Charles Darwin grew aware of that hard truth long before he published *The Origin*. By 1845 he had published his *The Voyage of the Beagle* and had established a fine reputation as a naturalist-adventurer. He was now concerned with the biggest subject, the biggest adventure of them all. He had already drafted a paper on evolution but did not submit it for publication; he knew that it would face the most horrendous opposition. He was seen as a brilliant amateur: a speculative paper from such a source would be regarded as mere fireworks. He wasn't really part of the club. He didn't have a solid body of work behind him. Now Darwin was a great writer of letters (which, younger readers may care to learn, are like emails only different) and they show at the same time his breadth and depth of mind and an apparently bottomless fund of generosity and modesty. Joseph Hooker, the great botanist, was his closest friend and most trusted colleague. They were exchanging views on a recently-published book that was bursting with bright ideas. Hooker wrote: "I am not inclined to take much for granted from anyone [who] treats the subject in his way and who does not know what it is to be a specific Naturalist himself."

Darwin was mortified. He wrote back: "How painful (to me) is your remark that no one has hardly a right to examine the question of species who has not minutely described many." Hooker was distressed that Darwin should have understood his words as a personal criticism, but Darwin was adamant that Hooker was right. He had a serious and terrible truth to unleash on the world: he needed to do it with an appropriate reputation for seriousness. He needed to show that his perfectly enormous and very general idea was backed up by detailed work on the very tiny and the very particular. He wanted to tell the world something about what species in general means: so he needed to devote himself to the questions posed by species in particular. So he spent the next eight years of his life (including two years off with illness) on the study of what he termed "my beloved barnacles".[*] His initial intention was to cover just a few species, but the more deeply he got involved, the more he realised that the entire group needed intense and detailed work. And that is precisely what he gave it.[†] A series of monographs revolutionised our understanding of barnacles. It also gave Darwin the street-cred he needed when at last the big book came out in 1859. It was a book written by a man who knew what a species was when he saw one.[‡]

Barnacles[§] have shells, like oysters and snails and cockles and mussels, so it seems obvious that they're all one and the same. But barnacles aren't molluscs: they are crustaceans, and so they belong in the phylum of arthropods. It's yet another of the counter-intuitive truths of the wild world: barnacles and limpets lie adjacent on the same exposed rock in the intertidal zone and seem to be living much the same sort of life. But they came at it from different evolutionary directions. Barnacles are more closely related to the

---

[*] Darwin was not the only one with a beloved barnacle in his life. In 1904 James Joyce eloped with Nora Barnacle. His father remarked: "Well, begod, she'll stick to him." They stayed together until Joyce died in 1941.

[†] His children, visiting friends, asked: "Where does your daddy keep his barnacles?"

[‡] See two excellent biographies: *Darwin*, by Adrian Desmond and James Moore, and *Charles Darwin* by John Bowlby.

[§] In medieval times it was generally understood that geese hatch from barnacles; or to be more precise, that barnacle geese hatched from goose barnacles, a group which do look a little like the heads and necks of geese. This somewhat odd notion made it acceptable to eat geese at times when meat was forbidden, like Fridays and during Lent. It was not, then, a story that people were overeager to prove false.

butterfly that strays across the beach than to the limpets that squat alongside them.

Barnacles are technically cirripeds, or curly-feet. There are about 1,220 species, and in their adult, static form they lie on their backs and extend their curly feet into the water above them and waft food particles down to be consumed. They are mostly to be found in shallow water and the areas between the tides, though some species are parasitic and live inside crabs. They go through two distinct larval phases, and that separates them from molluscs and allies them with arthropods.

This minute study of small but crucial differences in closely related species reveals without compromise the tendency of evolution to create radiations of species. Darwin perceived in delight how aspects of barnacles were analogous to the head of a crab but "wonderfully modified". Here is evolution in action: here is a clear example of the mutability of species. He observed further than the organ that worked as the oviduct – egg-laying implement – of a crab had become "modified and glandular and secretes a cement". The cement is used by the larva to attach itself to a rock and become static, or sessile, as scientists prefer to say. In other words an organ can change its function from one related species to another: a crucial truth about evolution.

Eventually, after those eight unending years, the damn thing was done. Thus the brilliant amateur was now recognised by all as a solid professional: a man who had done the hard yards. Thomas Huxley was to call the barnacle monograph "one of the most beautiful and complete anatomical and zoological monographs which has appeared in our time". By the end, though, Darwin wrote: "I hate a Barnacle as no man ever did before, not even a sailor in a slow-moving ship."* It was time to move on.

---

* Joyce also knew what it was to get fed up with a work of genius. "For seven years

Counter-factual history: Darwin decided against barnacles because it was too much like hard work, especially for a sick man.[*] He put together his big work without it: and it created no stir whatsoever. No one took it seriously. It was just some half-baked notion of a gentleman adventurer who thought a wealthy background and a best-seller were enough to make him a scientist. Darwin's name was forgotten. So: did the truth remain hidden for many years? Or did Alfred Russel Wallace persuade science that his own understanding of evolution by means of natural selection was valid? Would Wallace have written as convincingly, as exhaustively, on the subject as Darwin? Would he have been believed? Would this scientific and cultural breakthrough have been delayed for years had Darwin not decided to spend eight years looking at barnacles?

I've been working at this book – blast it," he once wrote of *Ulysses*. He then spent 17 years on *Finnegans Wake*.

[*] It has been speculated that Darwin suffered from Chagas disease, a parasitic infection that he might have picked up in South America while doing his stuff on the *Beagle*, though this idea has been losing ground in recent years.

Secret snakes

The Luangwa Valley is full of birds that eat snakes. The bateleur cruises the airways of the valley looking for snakes: the western banded snake eagle and brown snake eagle are specialists which mostly look for snakes from a high perched vantage point. The lovely lilac-breasted roller takes small snakes in the same way. You wonder how they ever find enough to survive, because I've hardly ever seen a snake in the valley. You don't often bump into one. Then there is the excellent *A Guide to the Snakes of the Luangwa Valley*, by my old friend Craig Doria, and Patrick Nyirenda. One of the spooky things about this book is that most of the sightings of snakes – and perhaps most especially the venomous ones – seem to have taken place in the camps and lodges. It's as if you're more likely to meet a snake while walking from your hut to the bar than you are out in the bush.

This doesn't represent the reality. It's just that there are more people-hours spent in camp than anywhere else in the national park. You can go looking for birds and they're everywhere to be seen. It's much harder to look for snakes. You are most likely to find one by chance, and the laws of chance state that you are most likely to see them in the place where you spend the most time. I remind myself of that when I take a pee in the middle of the night.

In other words, (1) snakes are more numerous than we think and (2) they are very talented at keeping out of the way of humans. Their shape gives them a different understanding of the world's possibilities: they are good at the low and the narrow and the hidden; they are also good at climbing in silence and obscurity. Being reptiles, they are very good indeed at being still. You can walk through a landscape of snakes without knowing it. In fact, I've seen people doing so: with their dogs merrily bounding about as they go.

Dunwich Forest in Suffolk is a favourite dog-walk, and is much crossed and recrossed. It's not a place where you're ever going to be lonely for long. Those interested in cold-blooded terrestrial beasts are herpetologists, herpers for short, and I was in the forest with a crack herper called John Baker. It was the beginning of spring. We walked along the south faces of the windrows, the long piles of brash and bracken. And time and again, he would say: "Adder." And I would ask where and he'd tell me and I still couldn't see it. Eventually I would almost prise it out of the landscape with my eyes: the same colour as the dead bracken fronds and the same pattern, but a warmer tone. And sometimes, very subtly, in movement. He must have shown me 30 or more in a couple of hours' walk. This is their place. They are venomous snakes, they are all over the forest, but it takes a person with the right eyes and the right educated mind to encounter one.

Few people are aware that England's green and pleasant land is home to snakes 5 feet long, 6 six, nearly 2 m. Female grass snakes can reach this length: and they have a respectable girth, too: substantial snakes. I saw one swimming in a pond in Suffolk: a majestic lady crossing the open water with impressive lateral undulations. I found it rather wonderful to think that even in safe southern England, there are snakes the length of hosepipes laterally undulating about the place, while in the most well-trodden of woods, venomous snakes get on with their lives without troubling a soul, and with scarcely a soul aware of their abundance.

I know this won't come as happy news to those who suffer from a fear of snakes – from ophidiophobia, to be more technical. So it serves me right that I must now move on to spiders.

The silk route

Silk is the glory of the spider. As Eskimos are (wrongly) supposed to have dozens of words for snow, spiders (genuinely) have many different forms of silk and many more different ways of using it. Building webs is just the beginning of it. I must tell you first of a thing of beauty and wonder that is all the work of spiders, something so lovely that it charms even me on a regular basis. In my previous place, I used to start most days with a climb up what we in East Anglia call a hill: certainly a reasonable slope by the most demanding standards. It used to catch the low morning sun in the autumn as I headed up to feed the horses. And when a bright morning coincided with a hefty dewfall, the field was transformed. It became an enchanted place, a fairyland of spiders. It seemed that every two blades of grass were linked by a golden thread a few inches long: a strand of precious metal that trembled slightly when the breeze shook it. It seemed as if I was walking through not a pasture for horses but a crop of gold: a waiting gilded harvest that had sprung in delicate strands from the bosom of the earth.

The sun, shining through the dew, had called them into being: a classic epiphany that is sometimes called a gossamer morning. Each thread represents the journey of a spiderling. A hatch-out of spiderlings continues with a great adventure, a surrender to chance. Each tiny spider extrudes several strands of silk, and these catch the air on a windy day, and up and away goes that spider to land – a few yards away, or sometimes a field of two away, or sometimes for miles, at times crossing continents and oceans. No island can spring up anew from the volcanic pulse of the earth without eventually being colonised by speculative spiderlings. The trick is called ballooning, though it's more like kiting, except that in a neat paradox, the string is used not for tethering but for flight itself. Ballooning spiders have been gathered as aerial plankton thousands of feet above the surface of the earth, and they have been found on ships in the middle of the ocean.

Spiders' webs are beautiful and admirable things, especially when the dew or frost chooses to set off their wonder. An orb-web – a roughly circular construction – of a few inches is a delightful thing to come across in the garden. They are rather more daunting when they reach over 3 feet, a metre across the diameter. The large woodland spider of Hong Kong is a particularly impressive craftsman – craftswoman, rather: the enormous black and yellow females stand in the centre, comfortably bigger than your hand with a web that looks as if it could snare a child. My copy of *A Colour Guide to Hong Kong Animals* has a photograph of a large woodland spider with a pipistrelle bat caught in the strands of the web. The bat is half-eaten:[*] she was saving the rest for later.

Roald Dahl wrote a short story called "The Visitor", in which the sinister and exquisite seducer Oswald Hendryks Cornelius keeps a greenhouse full of spiders; he harvests the webs every now and then and has the silk made into ties. He is the only person with the nerve to enter that dreadful place. Spider silk is remarkable stuff, and perhaps ever so slightly sinister to us humans because we can't empathise with the idea of producing it. How can you just *think* something into being: something so brilliant and so strong, so multifunctional? And like the man who could blow hot and cold from the same mouth, spiders can produce up to seven

---

[*] To be more accurate, half digested. Spiders don't munch and swallow: some inject their prey with digestive juices and then suck; others mince up their prey and then add the digestive enzymes. It's called external digestion.

different kinds of silk from the sa
make silk, but they can only mak
spiders who are the masters.

They use silk for catching prey,
of webs, most notably the wheel-s
also make traps for prey in the shap
and tunnels and tangles and domes.
lace; the bolas spider uses single strar
technique. Some spiders use silk to ir
ping it up like a mummy. Silk can be
with males making a silken parcel of s
dispersal by the ballooning spiderling
Webs are made of protein and can be
spiders will eat the webs of other spider
eat their own webs, as part of the econo
spiders will line a nest with silk. Some le
so they can find their way back to base
create silken drop-lines, so that when th
just let go and plummet – and yet do
into other possible dangers. Some spiders
silk that are triggered by passing prey. So
leaving a pheromone trail of silk.

It's extraordinarily tough stuff. The
strength and elasticity makes it astoni
Weight for weight it's as strong as steel all
it can stretch up to five times the length of
makes it incredibly hard to break. Silk is
in the way life works: when you get a ver
can be adapted and readapted again and ag
removed from its original propose. If silk ori
for a single way of catching prey (and that's
is now used in many different ways for that
also for life-saving, infant care, nutrition, esca
the world, and love.

Lizar
a rept
tratin
the ti
amor
snak
They
have
in th
geck
Out
larg
role
tio
hu
fai

wi
pr
th
tic
m
ch
th
ch
a
N
1

Disgusting clumsy lizards

ls are more or less the default reptile. Mostly, if it's ile, it's a lizard: around 5,600 species of them, pene- g all kinds of habitats from the Arctic Circle down to p of South America: up in the trees, on the ground, ng rocks, in the water, burrowing in the earth. Unlike es they have firm, locked jaws and external ears. mostly have four legs, but some of the burrowers done away with them, like slow worms, already met ese pages. Lizards range from tiny chameleons and os no longer than your finger to the komodo dragon. in the Indonesian archipelago on an island with no e carnivores, the lizard has taken the alpha predator and become a giant: up to 11 feet, 3.3 m, in excep- al individuals, and known on occasions to have killed nans. Most are less than 3 m or 10 feet, which is still a size.

Those who have lived in Southeast Asia are familiar h the house geckos shinning up the wall, operating on a vate theory of gravity, pop-eyed, with translucent skins it make most of their inner organs available for inspec- n. They are almost invariably welcomed because they eat osquitoes, which is an endearing trait, and have such a arming appearance, hiding behind framed pictures during e day and coming out at dusk. They utter a pleasing nuckle: in Bahasa Melayu (the Malay language) they are n onomatopoeic *cheechak*. Out in Africa you can find the Nile monitor, a seriously imposing lizard, which can make .6 m, 5 foot 3 inches, and weigh 15 kilos, 33 pounds. Like

snakes they taste the air with their tongues. They swim with immense confidence.

We associate lizards mostly with hot places, scurrying out of sight in an instant of time when your shadow falls across them in the heat of the Mediterranean day. To have the same experience, though at a lower temperature, is always a surprise in Britain, but the common lizard loves a good bask on a sunny day and is more often encountered than you might expect, though it is more frequently seen in the tail of your eye than in plain view.

Lizards are not to be underestimated in their adaptability. This is a lesson that Charles Darwin almost learned in the Galapagos archipelago during his famous trip on the *Beagle*. The Galapagos section of his great travel book is full of look-behindjer moments. When we read the book in the 21st century, we already know about evolution, we already know about Darwin on the *Beagle*, we know already about Gregor Mendel and how his pioneering study of genetics showed us the mechanism by which natural selection operates. We also know that the Galapagos are full of creatures that evolved uniquely for the place: and that as a demonstration of evolution in action, can't be bettered.

But the fascinating, almost bewildering fact is that Darwin didn't know that. When he found a unique lizard that dived into the sea to forage for seaweed, he had no idea that he was looking at one of the extraordinary Galapagos specialities: a creature that would become a classic example of evolution in action. Museum specimens of the marine iguana had been labelled with the erroneous information that they came from the South American mainland. Sometimes when you travel, you find something odd, but you think it's just you. You think the entire place is like that: you're just not used to it yet. So it was with Darwin and the marine iguana: he wrote about them as "disgusting clumsy lizards" and moved on.

But I suspect that a great deal of this bizarre stuff on

the Galapagos archipelago joined the great menagerie of sleeping influences on his future thinking. I suspect that the oddness, the uniqueness, stayed with him, and was part of the furtive development of his great notion: the one that finally reached its eureka moment when he turned to the work of Malthus, and read that more humans are born than survive into adulthood to produce children of their own. Yes: and that's true of all animals. So why do some survive and others not? Perhaps it's because some have an edge? And a nuclear bomb went off in his mind, later to be detonated across the entire world.

But there was one oddity that Darwin noted about the disgusting lizards. He picked one up and threw it into the sea. This did not discompose the lizard very much: it is a strong swimmer and made its way back to land – where Darwin seized it by the tail and hurled it out again. He repeated this again and again, the iguana always returning to the shore where Darwin was waiting. He wrote in *The Voyage of the Beagle*: "Perhaps this singular piece of apparent stupidity may be accounted for by the circumstance that this reptile has no enemy whatsoever on shore, whereas at sea it must often fall prey to the numerous sharks, hence, probably, urged by a fixed and hereditary instinct that the shore is its place of safety, whatever the emergency may be, it there takes refuge."*

Not stupidity, then. This was a perfectly adapted animal that failed to cope with radically changed circumstances. All the things that prepared it for life on the Galapagos were not enough to cope with something completely different. When such circumstances are reproduced on a large scale, extinction follows. Was this hurled and defiantly swimming marine iguana yet another of those sleeping influences?

---

* The passage is quoted by David Quammen in his excellent essay "The Flight of the Iguana", published in a collection of the same name.

The Kalahari Ferrari

What does it mean, not a true spider? The phrase baffled me as a child. Anything that ran about on eight legs seemed to express a very clear truth, and not a terribly welcome one. The idea of animals being untrue raised a small turmoil in my mind whenever I came up against it. Seals, I remember reading, had sharp teeth and lived by catching and eating things, but they weren't true carnivores (although of course they are now). Pandas, on the other hand, though they don't touch meat in the normal way of things, were true carnivores, and still are.

There are an awful lot of species out there that look pretty much like spiders and which belong in the class of arachnids, but are not spiders at all. There is a gulf between true spiders and untrue spiders. Arachnids also include quite a few species that don't look anything like spiders: creatures that look as if they should be classified quite differently. They all fit into the subphylum of chelicerates, which includes arachnids, along with the smaller groups of horseshoe crabs and sea spiders – which, I should add, are not all that closely related to true spiders. The point to a zoologist is that they all share a one-piece head/thorax combination. I trust I make myself obscure.* Perfectly? Then let us continue.

Some animals are capable of alarming you because they can kill you, others because they give you the creeps, still

---

* As Sir Thomas More says in Robert Bolt's *A Man For All Seasons*. Do something as a school play and it's likely to stay with you for life, which is a good argument for schoolteachers choosing only the greatest plays.

others because they make you jump. Mice and bats fall into the last category, but by far the best are the sun spiders or solifugids. Latin for fleers from the sun. Not, of course, true spiders. They share a common ancestor with spiders but have diverged to form a separate group: I put in that clarification for my childhood self. One difference is that most spiders do a lot of hanging about. Solifugids don't. As the sun goes down they get going. Instead of waiting to see what the world will bring them in the way of food, they set off at top speed until they bump into something edible. They tend to have a preferred racetrack, and they go round it again and again. They can reach speeds of 16 kph, 10 mph, which seems at least twice as fast in a confined space. When this space is somebody's living room and you're sitting in it having a civilised conversation, this can be rather startling. Solifugids have a fine knack of making you think they have given up and gone away before reappearing at breakneck speed, making you – all right, making me – emit a girlish squeak and pour a tablespoon or two of cold beer onto my crotch. They have a great nickname: the Kalahari Ferrari.

Harvestmen, another group of arachnids that don't count as true spiders, are disturbing in a quite different way. They have incredibly long, thin legs, horrifyingly fragile things of cobweb thinness which support in the middle an unnaturally tiny body: or to be more accurate, an unnaturally tiny abdomen plus head/thorax combo. They appear over baths, looking as if they're about to fall apart and drop bits of themselves all over you*... but perhaps I am telling you more about myself than harvestmen here. Arachnids comprise around 65,000 described species. A lot of them are tiny, but still look pretty alarming under magnification. Mites are many and various, some feeders on detritus, some

---

* And as a matter of fact, they often do precisely that. They make up the order Opiliones, which comes from the Greek for shepherd, so named because shepherds in some places performed their duties on stilts.

predators, some parasites. They include ticks, which feed on blood.

Scorpions are also arachnids. There is a glorious set-piece in one of Gerald Durrell's books, in which Gerry's brother Larry opens a matchbox to light a cigarette and finds, instead of matches, a scorpion and young: "It's that bloody boy... he'll kill the lot of us... look at the table... knee-deep in scorpions..." Every morning when I am in the bush I bash out my boots before putting them on, to shake out the scorpions. I've never found one yet: not a detail that will stop me banging. There are a couple of thousand species of scorpions, and the champion* is 20 cm or 8 inches long. I'm pleased to tell you that "only" 30 or 40 species carry venom strong enough to kill a human. Even scorpions are wary of scorpions. When they court, they seize each other by the pincers, so each can dissuade the other from turning an amorous encounter into a cannibalistic feast. When I was at Sunnyhill Junior School we were shown a Walt Disney film of this behaviour. It came, alas, with comic music, as if the scorpions were doing a silly dance to amuse humans: the darkened classroom was filled with silly glee. I still remember my hot-eyed fury at this. I couldn't bear the thought that people thought animals silly. Surely if anyone was silly here, it was humans. No animal has been put on earth to amuse humans: to think so is demeaning to my family and to other animals. True spiders, untrue spiders and all their fellow arachnids may give me the jumps, but that's my folly and failure.

---

* *Hadogenes troglodytes*, the South African rock scorpion, or flat rock scorpion. Only mildly venomous.

Good luck, little metaphor

Tiny, and swimming hard between my finger and thumb as I picked him up from a plastic washing-up bowl, away from his brothers and sisters. All of them were flippering and flappering as if their lives depended on it, and for the best of reasons. Tenderly, I placed him on the black sand of the beach: unhesitatingly he set off. Towards the sea, not away from it: the compass sense that will guide his swimming sisters back to the same beach to lay eggs of their own drives the young turtle towards his destiny in the waves. I set another four on the same trail; soon here was a line of turtles going through a ritual not unlike birth: the transition from the allwombing* beach to the perilous and enthralling sea.

I was on the Pacific coast of Colombia where the hot, wet forest comes right down to the sea and the humpback whales rise and breach in the waters within sight of the shore. I was visiting a project for the olive ridley sea turtle. Their eggs are gathered from the nests on the beach and kept safe: safe from hungry humans, safe from feral dogs, safe, too, from other turtles, who come onto a crowded nesting beach and accidentally dig up each other's nests. At the project headquarters the eggs are watched and monitored and when hatched, the turtlings are released with the sea before them. That way close to 100 per cent of the hatchlings reach the sea: that's already far better than anything that could be hoped for in the crowded modern world. But

---

* A Joycean coining.

it's still a dismaying sight: these tiny creatures setting off so optimistically into so cruel a world – well, it rather breaks your heart. I might have got us all drowned, or at least stuck, because I refused to leave the site of the release until my – that should be "my" – turtles had reached the sea. The driver of the waiting vehicle had to perform minor heroics to beat the incoming tide and get us to our destination. But no one can walk away from a seaward-scrabbling turtle.

Television programmes have made this an enduring image of the vulnerability of nature: turtles forever striving for the sea. It has become an image of blind hope within the hopelessness of life: many boys seek to become professional footballers and as few make it to the top as turtles make it to adulthood. It's also an example of the way humans take advantage of weakness: a trusting turtle lays a cache of eggs, only for the hard work to be undone by endlessly hungry people. You see turtles walking towards the sea and you say: this can't possibly work.

Of course, if you take out the unsustainable harvesting of the eggs by humans, the strategy works very well indeed and has done for millions of years, since long before humans were around, since long before mammals had much of a say in what happens on this planet. A female can live 20 years and more, each year laying one, two, or even three batches of eggs, each one of more than 100. Conservatively, that's 3,000 eggs: 3,000 spins of the roulette wheel. If only one of these survives to adulthood and breeds in its turn, she's fulfilled her minimum biological ambitions; the same rule works for the males. In a pre-human world that's pretty decent odds, as the survival of the species makes clear. It's only in the last century that things have changed.

A female might be covered by eight males for each batch. Each male is capable of fertilising a number of eggs in the same clutch, while the sperm itself competes within the female. Olive ridley sea turtles are not giants, but they're capable of reaching a decent size, the females up to 46 kg, more than 100 pounds, and 70 cm or 2 feet 3 inches in length.

Turtles, terrapins and tortoises make up the order of Testudines: an easy-to-follow bit of taxonomy, since they all dwell in a private armoured car (or ironclad water vessel if you prefer): a carapace above, a plastron below. There are about 300 species of them in 14 families. The giant tortoises of Galapagos may have been another of those sleeping influences on Darwin: certainly, they provide another of those look-behindjer moments in *The Voyage of the Beagle*. Darwin, being a young naturalist, accepted what he had been told: that the tortoises were exotics, brought there and dumped by humans. The prisoners from the penal colony there said that each island had a different tortoise: the vice-governor said that he could tell a tortoise's island of origin from one glance at the shell. Darwin could have made a hefty collection of tortoise-shells: nothing easier, since there were many lying about, often used as flowerpots after the animals had been caught and eaten by humans. But he moved on: and those who visit the Galapagos can look at the tortoises and understand them as a classic demonstration of the workings of the principle that we now call Darwinism.*

---

* Darwin did, however, notice the mockingbirds of the Galapagos. He realised that the mockingbirds on Charles Island differed significantly from those on Chatham. So he labelled his mockingbirds island by island – and it was the mockingbirds, and not the more famous finches, or the iguanas or the tortoises, that unlocked the mystery of life. Darwin's revelation was itself mocked without mercy, or for that matter thought, before it obtained scientific acceptance.

10. OLIVE RIDLEY SEA TURTLE

Twenty centimetres!

When I first arrived on Lamma Island, I was naturally greeted with horror stories about the terrifying wild creatures that would visit my ground-floor flat. Being even then something of a traveller, I laughed them off. And I was right to do so: the hard-bitten hard-biting centipede that had been so graphically described turned out to be a millipede: no rapacious carnivore but a charming vegetarian, bumbling across the floor with the air of an absent-minded Tube train that had forgotten its tunnel. I gently removed it from the house in the dust-pan: it even had the courtesy to curl up like a liquorice allsort as I did so.

And then, of course, the real centipedes started arriving: large, fast, fierce and wriggling hard. These are impressive and alarming, and they can bite. Everybody on the island had a centipede story or two. I was awoken at night by a tinkling sound from the electric fan above my head. We argued: something's there no there isn't yes there is no there isn't go and look there's no point turn the light on oh ALL RIGHT. As I did so a 6-inch-long centipede fell from the safety cage surrounding the fan to land on my pillow. A nip from one of these things can hurt: my neighbour Sam was bitten on the balls and even he, a person of infinite good humour, found it hard to laugh that one off. Bloody funny for the rest of us, of course, and hardly the worst thing he took to his bed that year, as we all heartlessly agreed.

Then there was a pretty and charming Frenchwoman who lived on the island for a few months. She too was troubled by a nocturnal invader, and she made a terrified small-hours phone call of distress to Chris. Chris solved everybody's problems on the island: he was that sort of man. Just how many problems he solved that night must remain the subject of speculation, but the French lady contributed one memorable phrase to the story. She was much moved by the whole encounter and was inclined to tell the tale again and again, always returning to the same phrase: "It was twentee centee-meetairs!" Twenty centimetres!

The unsporting *A Colour Guide to Hong Kong Animals* gives a maximum length of 13 cm to what it calls the urban giant centipede. I have found other reports of the same species, now called more simply the giant centipede,* up to 8 inches. But if you seek a giant worthy of the name, look for *Scolopendra gigantea*, the Peruvian, or sometimes the Amazonian, giant centipede. It can reach a foot and more – thirtee centee-meetairs! – and preys readily on vertebrates.

Centipedes and millipedes are all part of the same group of myriapods, which contains 13,000 species. They all have segmented bodies: millipedes' segments are fused in pairs so that they appear to have two pairs of legs to each segment; centipedes very clearly have but one pair per segment. Millipedes are mostly found in wet forest where they eat decaying plant matter and play an important role in the commodius vicus of recirculation that keeps the rainforest forever self-renewing. None has a thousand legs: *Illacme plenipes*, already mentioned in these pages, is the champion with 750 legs. Some species have fewer than ten legs, which seems hardly fair. The pill and box millipedes have

---

* Common names, or vernacular names, are a constant source of confusion, and people are always making doomed efforts to tidy them up and rationalise them. Some people insist that we call the robin the European robin, and that we call the bearded tit the bearded reedling, because it's not closely related to tits. I rather enjoy the chaos myself: it's part of the great Babel of world language. If you want precision, you have the designedly unambiguous scientific names. At least, that's the idea, but here the names are constantly changing in response to new discoveries. The point is that common names reflect intuitive taxonomies rather than scientific ones. As such, they are as valid in their way as organised scientific taxonomy.

abandoned the Tube-train body plan and contracted to become short and bulbous.

Myriapods are fine beasts and worth a chapter. But now we must leave them and every other class of rthropod and every other phylum of invertebrate alone, for it is time to enter the last class of the inverts. And I have saved the greatest for last.

Shape-shifters

I have perpetrated a gross libel on Sunnyhill Junior School, Streatham. I can clearly remember one incident in which I received an important bit of education from that establishment. It was the time when we did frogspawn. There in the classroom was a glass tank containing the black-blobbed jelly that was awaiting its destiny. It was strange: as if the other 44 members of the class were to be given a guided tour of my world. Not that I had ever seen the miracle associated with frogspawn for myself, for in those days most of my engagement with wildlife was platonic. I read about it, I thought about it, I was deeply in love with the idea of it: but all without physical reality. It was, back then, an affair of the mind. Which made the time of the frogspawn especially vivid.

The spawn hatched out into tadpoles. How unlike a frog! These tubby little black torpedoes looked as if nothing could be further from their minds than hopping, leaping, catching flies and going *ribbit-ribbit*.* They were fish with destiny: destiny's larvae. They swam about and breathed through gills, and if you were to drag them out of their tank into the deadly air of the classroom, they would drown

---

* Of the nearly 5,000 or so species of frog that exist around the world, only one, the Pacific tree frog, says *ribbit-ribbit*. It can be found in California. Frogs have a glorious and expansive range of sounds, and *ribbit-ribbit* is but one. It has become accepted as the archetypal frog sound because of Hollywood. When early talkies required atmospheric outdoor sounds they stuck a microphone outside the studio to catch the locals, regardless of where the film was actually set. Suddenly, all the frogs in the world had but a single sound. Rather like all humans in the world speaking American.

as a fish drowns. They had no legs: they swam with their tails. They fed on algae and plant matter. But gradually the changes came: the emergence of back legs, which in a day or two were used for propulsion. The tail began its retreat, the front legs appeared: and then they were not tadpoles at all; just minutely perfect froglings ready to start a life on land, living not on plants but insects, breathing not through gills but lungs.

Seeing this for myself was not like reading about it. Seeing this transition made me wonder. Here was an animal that turned into a completely different animal: vegetarian* to carnivore, water-breather to air-breather, legless swimmer to four-limbed leaper, a creature restricted to water to one with the great solid world before it. I wondered if I was a kind of larva too, if I would be radically different when I grew up. But no, that wasn't the case with humans. I could see that it was only a difference in degree, in size, in proportion, in strength, in mental power, in sexual awareness. I was not a fish that would become a man: I was a boy already timidly growing towards manhood. Not the same thing at all. What was it like, then, for a tadpole, its mind circumscribed utterly by its need to be in water, to be suddenly confronted with the vast impossibilities of land? Did it take it all in its stride and hop boldly on as if it were the most natural thing in the world? Or did it feel a strange twinge of apprehension, a feeling of, this can't be right, can it?

Not the most scientific approach, it's true. Perhaps I knew that at the time. But all the same, it was an appropriate response to the impossible fact of metamorphosis: shape-shifting: moving from one state of being to another. Oh yes, sure, a great metaphor for human development in a bildungsroman sense of the term, but as a real, happening non-metaphorical thing its impact was much greater. Your

---

* Tadpoles aren't exclusively vegetarian; in fact, they will happily take to cannibalism when there's an opportunity – like, for example, a great crowd of them in an aquarium.

scientist landing from Mars or Tralfamadore* to inspect the ten million or more aliens he came across would surely look at the tadpole and cast it in with the fishes, and then put the frog somewhere completely different. They were different animals, with different body plans who lived in different ecosystems: how could they possibly have anything to do with each other?

Our mammalian young are at least roughly recognisable as smaller versions of adults. Some of us leap almost at once into something alarmingly like maturity; I once assisted at the birth of a foal and within an hour it was standing upright and suckling. Others take longer, but young animals become old animals in a continuous unpunctuated process. Some birds (the altricial species as mentioned before) start off looking reptilian and fluffy, but they shed their down and grow their flight feathers and do so without abrupt transition. Reptiles mostly hatch out as perfect miniature versions of adults: the same proportions scaled down. Baby crocs are no cuter than the adults on the Grameti River. But now we have reached the amphibians we must get our heads around the idea of metamorphosis. We'll meet it again in the larval forms of fish, we have come across it many times over in our circle through the invertebrate phyla, and we will encounter it again as we move into the final class of the inverts. It's a frequent matter-of-fact and quite unextraordinary thing: just one that is light years remote to us as shape-keeping mammals.

---

* Tralfamadore is a colossally distant planet that crops up in the writing of Kurt Vonnegut, especially *Slaughterhouse 5*. Tralfamadorians exist simultaneously across all of time. How does a novel work, then? "Each clump of symbols is a brief, urgent message – describing a situation, a scene. We Tralfamadorians read them all at once, not one after the other. There isn't any particular relationship between all the messages, except that the author has chosen them carefully, so that, when seen all at once, they produce an image of life that is beautiful and surprising and deep. There are no beginnings, no middle, no end, no suspense, no moral, no causes, no effects. What we love in our books are the depths of many marvellous movements seen all at one time." Perhaps what I am attempting with this book is to create a Tralfamadorian novel.

That transformation appears to us as one of the greatest miracles of life after life itself. Why does it happen that way? It seems incomprehensible that so complex a system, so Byzantine a route towards adulthood, should have evolved. There is a sneaking feeling in the human mind that evolution should be a process of refinement: here is evolution as willed complexity, as deliberate over-complication. The only justification is that it works, and has worked for endless ages. Amphibians pre-date reptiles, and before reptiles led the way across the land with their water-tight skins and their impermeable eggs, amphibians were the biggest creatures on the planet, with Prionosuchus reaching the astonishing length of 9 m, 30 feet.

Amphibians are all pretty small now, at least in comparison with the extinct monsters. There are about 7,000 species living: the fourth of the traditional classes of vertebrates, squeezed uncomfortably between reptiles and fish. Most of them go in for metamorphosis, though a few species have evolved a less complex and less vulnerable strategy. And they embody a spectacular paradox: their evolutionary triumph was to escape from water but their essential nature compels them to be forever dependent on the ready availability of moisture. When an amphibian gets too dry it dies of desiccation. Its life is dependent on dampness. Amphibians leave the water, but they can't afford to let the water leave them.

Second innings

"He played that stroke like an old lady poking her umbrella at a wasp's nest... the umpire signals a bye with the air of a weary stork... that was the stroke of a man knocking a thistle-top with a walking stick... a late cut so late as to be positively posthumous... Nine runs off the over, 28 Boycott, 15 Gower, 69 for two. And after Trevor Bailey it will be Christopher Martin-Jenkins."[*]

And so, to the soothing rhythms of *Test Match Special*, BBC Radio's ball-by-ball commentary on Test match cricket, with the great John Arlott always ready to paint a word-picture of the distant struggle, a beautiful butterfly's future was safeguarded in the West of England. Jeremy Thomas[†] did much of the research in the six summers between 1972 and 1977, hour after hour and day after day; *TMS* helped him to keep going. Thus the mysteries of the large blue butterfly's life were understood in enough depth to allow an extraordinary conservation operation to be put in place. The butterfly went extinct in this country, but in 1983 it was reintroduced and it has been here ever since. It's one of the great secret triumphs of conservation, and it all comes down to Thomas's work in the field.

The large blue needs a lot of understanding. It begins its life as an egg planted on wild thyme. The adults live only for a few days: as many as two years have been spent preparing

---

[*] Last words of John Arlott's final broadcast on *Test March Special*, on September 2, 1980.

[†] *The Butterflies of Britain and Ireland*, by Jeremy Thomas, illustrated by Richard Lewington, is one of the essential wildlife books.

for the supreme moment: the mating, the egg-laying, the almost immediate death. Mick Travis, the schoolboy rebel in the film *If...*, gives his full attention to a gorgeous pin-up: "Only one thing you can do to a girl like this. Walk naked into the sea together as the sun sets. Make love once. Then die." A lot of insect life is exactly like that, though seldom is the creature as gorgeous as a large blue.

The egg hatches: a small caterpillar emerges and begins to feed on the pollen and seeds of the wild thyme. It is also perfectly prepared to feed on other large blue caterpillars: they are cheerfully cannibalistic, and so in the end, each flower head has but a single victorious and hard-munching caterpillar. After two or three weeks, the caterpillar is ready for the next stage. It is not much larger but it has developed the organs that will allow it to move on. Chief among these is a honey-gland that can secrete sweet drops of moisture. At this stage, the caterpillar drops to the ground of an early evening and hides – hides hoping to be found. With luck, the caterpillar is found by red ants from the genus *Mymirca*. Plan A is to be caught by the species *M. sabuleti*, which gives the caterpillar the best possible chance.

The ants that find the caterpillar start to tap on it, making it release the dew. Soon, they're all over it. This is a process that takes four hours. At this stage the caterpillar rears up: now smelling strongly of the ants, and therefore of the ant's nest. At this, the ants become convinced that the caterpillar is one of their own grubs that has, for some unexplained reason, strayed from the nest. Filled with sudden anxiety, they hustle it underground as fast as possible. There are some larvae of other species that enter an ant's nest and establish a wholly benign relationship with its hospitable inhabitants. They offer honeydew and get the protection of the colony in return. Not so here. Once underground, the caterpillar behaves rather like a cuckoo, but even more damagingly, if anything. Its first act is to attack and eat a grub. And that's just the start. It eventually turns into a bloated white maggot that dwarfs both the solicitous ants and their helpless grubs. It hides in the colony and hibernates. When the warm weather comes again, it mostly lies doggo, making occasional binge-feeding raids on the grubs. It might spend as much as two years underground, feeding in this way, during which time it will eat an estimated 1,200 grubs. Not that the billet is as cushy as it sounds: sometimes the ants attack the large blue larvae, in the belief that they are rival queens; sometimes, when there is more than one large blue caterpillar in a nest, they can be out-competed and die of starvation. But a successful caterpillar will then pupate, later to emerge as a butterfly. The only problem is that it is trapped at the bottom of an ant colony.

Ants can make sounds by means of friction, a process called stridulation. The butterfly can mimic this. It can sound like a queen ant, stirring the workers into admiring service. Lovingly, they escort the intruder to the open world: not to fly forth and establish a new colony of ants, but to try and raise caterpillars that will prey on a colony of ants in their turn. The adult large blue emerges and inflates its wings, and after 45 minutes or so, sets off in a frantic search for sex, reproduction and death.

As a result of Thomas's work, the crucial importance of *M. sabuleti* was established, and so were the needs of the ants themselves. For all that they are the aggressors rather than the victims, the butterflies can't live without the ants: it is the ants that control the numbers of large blues. The grass in the traditional large blue colonies had been allowed to grow much longer than in the past, because of changes in the numbers of grazing animals. That meant conditions had become too cool for the ants to make a living. When a new grazing regimen was established, the ant numbers recovered and they were soon ready to receive their honoured and ungrateful guests.

And so the cycle continues: a bewilderingly complex life to begin our journey through the bewilderingly complex world of the insects. Insects are implausibly complex: often enough in their lifestyles, and more even than that, in the mind-addling madness of their numbers.

When I was a rain god

"And as he drove on, the rainclouds dragged down the sky after him, for, though he did not know it, Rob McKenna was a Rain God. All he knew was that his working days were miserable and he had a succession of lousy holidays. All the clouds knew was that they loved him and wanted to be near him, to cherish him, and to water him." Rob McKenna is a lorry driver in Douglas Adams's *So Long and Thanks for All the Fish*, in the Hitchhiker series. There have been occasions when I wondered if I wasn't something of a rain god myself. There was the time I went looking for Dupont's lark in the driest part of the driest part of the Spanish Steppes, which makes it the driest place in Western Europe. And it pissed down and we got quite tremendously stuck. We had been admiring the fragile resilience of the ancient shin-high vegetation: we ended up ripping out great armfuls of the stuff and cramming it under the drive-wheels. And then there was the time I went to the Kalahari, not to look for the Ferrari, but to see the desert in action. As soon as it knew I was coming, the Kalahari had its biggest rainfall for ten years. My desert bird list has five species of duck, two of grebe and one of pelican: not a bad haul for one of the driest places on earth. And also there were plenty of African fish eagles: birds I know well from the Luangwa Valley, where there is a pair for every mile or so of river and for every permanent lagoon. They live by plunging into the water to grasp fish. They are birds of lush, green, wet places. The desert was full of them.

The reason they were there was as great a miracle as any wrought by a rain god. They were paying a visit to exploit a sudden bonanza of frogs. They had become African frog eagles and patrolled the suddenly dampened desert, gorging on frogs. For the Kalahari has regiments of frogs hidden beneath the dry unforgiving soil. They appear during the shatteringly brief wet periods. They hatch from eggs laid in ephemeral pools and puddles, rapidly go through their metamorphosis, become adults, and then, as the desert dries up again, they disappear beneath its surface and wait. They seal themselves off, wrapping themselves up in a kind of Clingfilm, which means they don't dry up. And there they wait, for years if necessary, until the kiss of the rain turns them into princes, or at least into living frogs again. How awful it is to contemplate the lot of a frog that has waited in his parcel for five torpid years, emerging at last into a briefly green and briefly pleasant land – only for an African frog eagle to seize him before he has made a single hop. But that doesn't mean the system doesn't work: enough frogs can survive, take advantage of this brief wet window to mate and lay eggs before they die: eggs which hatch out and become the next generation of prepackaged desert frogs. It's not a waste: it's a scheme that works for many species. After all, how many acorns does an oak tree produce?* But sometimes – well, rather often, actually – life operates on an economy that is alien to us humans. It's not wasteful; it's weird.

All of which shows the amazing ability of life to fill some of the least likely niches on earth. Desert frog should be as much of a contradiction as flying penguin, but it's an exact description. It's just one of the ways in which frogs manage to make a living: around 4,800 species in 33 families, ranging in size from *Paedophryne amanauensis* of Papua New Guinea, which measures 7.7 mm, 0.3 inches,

---

* That's been estimated at 13.5 million.

to the Goliath frog, *Conraua goliath*, of Cameroon, which can be a foot long, 30 cm, and weigh 3 kg, 8 pounds.

Zoologists make no distinction between frogs and toads:* they're all categorized as frogs, some of which have the common name of toad. Within the class is the family of Bufonidae, or "true toads": but that's confusing too, because they're not all warty and land-loving. There are warty frogs outside the Bufonidae, and within it, smooth-skinned and highly aquatic animals. It's just one more example of the fact that wherever you find life you find confusion, just as you find confusion on every page of *Finnegans Wake*. There are ways of understanding both, along with plenty of moments of dramatic insight and sudden wonder: but you never think you've got the problem licked, or even much of it.

Frogs sing, and in the most unexpected ways. They are the night soundtrack of everybody's adventures, the lullaby after the excesses of the day. When the big rains hit Lamma Island the valley would be roaring with bull-frogs. In Luangwa the nights, so full of threats and dangers from enormous mammals, echo prettily to the sounds of the painted reed frogs, tinkling from the vegetation around the water's edge like wind-chimes. These days everybody likes frogs: high, protuberant eyes, wide mouths, powerful back legs. But Linnaeus wrote: "These foul and loath-some animals are abhorrent because of their cold body, pale colour, cartilaginous skeleton, filthy skin, fierce aspect, calculating eye, offensive smell, harsh voice, squalid habitation, and terrible venom." So perhaps the popularity of frogs is a modern thing, a response to our increasing separation from the non-human part of the Animal Kingdom, and our need to keep hold of it.

David Attenborough's brilliant television series *Life on*

---

* In fact, the distinction only really works in England, where in historic times there was one (or some now argue two) species of frog, and two very similar species of toad – so one name for each was a no-brainer. But it's not an idea that exports.

Earth, of 1979 – I saw it in Hong Kong a year later – changed many people's understanding of the world we live in. And it was the frogs that turned out to be the stars: the frogs that provided the revelation that life is more wonderful than we are capable of imagining. The programme's enduring image, the cover-girl,* as it were, was the red-eyed tree frog of Central America; I went looking for them in Belize but failed to get lucky. And the sequence that nailed the impossible wonders of life on earth for all time was Darwin's frog, which for the first time was filmed giving birth.

Giving birth? Surely, not, you say, for I've just been going on for some pages about the way frogs lay eggs and go in for metamorphoses. But one of the problems of laying eggs in a small body of water is that they're hard to hide and there can be many creatures looking for them. A tasty unmoving morsel won't survive long. So some frogs have gone in for parental care: and adopted some of the most extraordinary tactics in the Animal Kingdom. The male midwife toad carries eggs in strands around his hind legs; the pouched frog carries tadpoles in pouches on his sides; the tungara frog builds a nest of foam; the Surinam toad raises young in pores on his back.

The male Darwin's frog takes the tadpoles into his mouth and keeps them safe in his vocal sac as they develop: thus the organ he used for wooing and winning their mother becomes their nursery. And when they have gone through their metamorphosis and are ready to face the world, they leap from his mouth in a strange mock-birth.

---

* It was Attenborough himself who took the picture – is there nothing the man can't do well? He's also a very decent pianist.

Laser epiphany

What's the most dangerous animal you've ever got close to? That's easy, it's an insect. What's the most beautiful animal you've ever seen? That's a bit harder, but it's still probably an insect. What animal has caused you most pain? Insects, without a doubt. What animals have brought you closest to despair? Insects again. Insects: there's no end to them.

So let's start with a lecture on ecotourism. I'm sympathetic: one of the good things we can do for the wild world is value it, and one of the best ways to value it is to go out and see it. But the lecture was taking place in a lodge that looked out over the Pacific coastal rainforest, for we're back in Colombia here, and the world beyond the lecturer's shoulder was full of the most excellent distractions. And all at once, for a period of time that was a little less than a second, I was overwhelmed by beauty. It was a sudden shaft of blue light: bluer than all the kingfishers in the world, colour in the form of pure light mainlined directly into my brain, apparently bypassing my eyes completely. And then gone, completely vanished, not a trace remaining. One of the great traditions of Italian football supporters is to try and blind opposing players, especially the goalkeeper, with a laser pen. I felt as if some hidden member of the *tifosi* from across the jungle clearing had lasered me. I knew what it had to be – could only be – but I was baffled by its instant disappearance. So perhaps I had imagined it. I stared out slightly to the right of the speaker's talking head with a deeply interested and concerned expression on my face and after a while – pow, it happened again. But for no longer than before: slightly less, if anything.

A couple of days later, up in the Andes around Medellín, I had a much better look. A full two seconds, perhaps even longer. And it was unmistakably an insect this time, a butterfly, of course, a blue morpho, to be precise: and perhaps the loveliest thing I had ever set eyes on, the lovelier for the tantalising brevity of the experience. Its upper wings are a fierce sunstruck metallic blue: but the underwings are dull and drab and cryptic. As it closes its wings or changes the angle of its flight it vanishes from sight.

That's beauty: what about danger? Mosquitoes* kill more people than lions and elephants and hippos combined; the tsetse fly, which gives a human a bite of piercing pain that lasts about half a second, has devastated communities because of the cattle disease it can carry, and also the sleeping sickness it can impart to humans. "God bless the tsetse fly," old Norman Carr, the founding father of Zambian conservation, used to say. "Without them we'd all be cattle-ranchers." The tsetses make the valley a no-go area for domestic cattle, so the place has stayed wild: a story true of many of Africa's great national parks. Insects have caused more devastation to humans than any other class of animals on the planet. The Colorado beetle has destroyed potato crops, and for every crop that humans have ever grown there are insects trying to get in the way. The rat flea has carried the plagues that shaped history. On the other hand, the pollinating services of insects make life on earth possible for humans. Let's return to the concept of "ecoservices", a notion that has emerged in recent years, according to which people put a cash price on the ways in which the wild world materially benefits humankind. It has been calculated that annual sales that depend on pollination

---

* The interior decoration on Lamma Island was based around the repeating motif of a flip-flop outline with a splatter of blood in the middle.

by animals add up to US$1 trillion; that annual services provided to farms by animal pollinators are worth US$190 billion; that two-thirds of major crops grown by humans rely on animal pollinators.*

The word insect means cut up. Insects are divided into three: head, thorax and abdomen. Each (mature) insect has six legs, and a chitinous exoskeleton, one pair of antennae and compound eyes. It's a body plan that seems almost infinitely adaptable: around one million species of insects have already been described, and as for the number still awaiting description, you can pick your own order of magnitude, because it's as good a guess as anybody else's. Insects exist in vast abundance and in impossible diversity. In every way, insects are many. More than any other group of living organisms on the planet, they are the driving force of the earth and the creatures that live on it.

And so we call them bugs. I hate that. Not because it's an unnecessary Americanism, but because it's an expression of dislike and contempt, and it alienates us yet further from the planet we live on. A bug is something that bugs you and is best kept out by a bug screen or demolished with bug spray. A bug in your computer programme is by definition a bad thing. A bug is everything bad: and its catch-all name is given to all small invertebrates. The (quite excellent) British conservation charity for invertebrates is called Buglife: perhaps even its executives and trustees wince a little at the name, for all that they know that a conservation charity for invertebrates has got to punch way above its weight if it wants to be heard, so a catchy name matters. But all the same, it offends me on a bug's behalf that all insects are by definition bad. Is the blue morpho a bug?

Insects are the only class of animals that have created an entire industry aimed at trying to kill them. It is an industry that, in the 1960s, came close to destroying the wild world,

---

* From Tony Juniper's splendidly readable *What Has Nature Ever Done For Us?*

as documented in Rachel Carson's world-changing book *Silent Spring*. The human response to that book was a clear indication of the fact that we have a great desire for the wild world to survive – for its own sake, not just because we depend on it for our existence. Insects can change an ecosystem: the ways in which humans seek to control them has changed the world, to the extent that the falling number of insects is beginning to be a serious concern. I was taught, and so were most of us, that the Age of Amphibians was followed by the Age of Reptiles, which was itself followed by the Age of Mammals, which became the Age of Man.* But ever since flowering plants spread across the surface of the earth, it's been the Age of Insects. They'll outlast us.

---

* The Age of Woman was for some reason never mentioned.

Death by frog

The golden poison frog is probably the most poisonous creature on earth. Chickens and dogs have died from contact with a paper towel on which this frog has walked. It's found only in Colombia, and it's tiny: a little scrap of charming, elegant, yellow death. Its skin is drenched in batrachotoxins, alkaloid poisons that prevent nerves from transmitting impulses. Come into contact with the stuff and you die of heart failure. Not that the frog is particularly anxious to come into contact with you: the poison is entirely for self-defence. The frogs are a decent size for poison arrow frogs, reaching 55 mm or a couple of inches in length, and they carry a single milligram of poison. This is enough to kill 10,000 mice, 20 humans or two African bull elephants. Death comes within minutes.

Inevitably, these fearsome creatures play a significant role in human cultures. The Choco Emberà Indians use them to poison the darts they use for hunting. Gently brush the tip of your dart or arrow on the back of a living frog; you can and should do this without harming the frog. You then have a weapon that will be deadly with a mere scratch, and will remain so for two years. Word of advice: handle these weapons carefully.

Because of all this history, this danger, this charisma, these frogs have become much desired by frog collectors. It's dangerous for an animal to be greatly hated by humans, but it can be just as dangerous to be greatly loved. The best survival strategy is to be ignored. The golden poison frog has suffered because its loving collectors have taken it from the wild without any thought for how the it will survive. Many populations have died out as a result. The frogs lives in primary rainforest in Colombia, and so inevitably, they are also suffering from the destruction of their habitat. They have a rather patchy distribution across an area of less than 250 square kilometres of rainforest on the Pacific coastal plain: it's a lot less than it sounds and it doesn't sound much to start with.

This frog has also suffered from one of the great contradictions of wildlife: peace is a bad thing. The area in which the frogs live is a great deal more secure than it has been in recent years; as a result, people are moving about more freely. Gold-miners are moving in, illicit coca cultivation is on the rise and timber companies are likely to arrive at any second. All this is the worst of news for golden poison frogs: you can't, alas, poison a bulldozer, not even with a paper towel. We have reached a state in which the animal that has developed the most effective method of protection in the entire history of the earth is now dismayingly vulnerable. It is completely unprotected by law.

In 2010 the Colombian conservation organisation ProAves launched a series of expeditions in search of remaining populations of the frog. They found to their dismay that it was no longer on any of its known traditional sites; it took extraordinary exertions before five tiny populations were identified. ProAves is now in the process of identifying the right areas to create the first protected zones for the golden poison frog. It is doing this with the support of the World Land Trust, a partner organisation in Britain that has agreed to fund the land purchase.

Astonishing: we are on the edge of wiping out one of the most extraordinary and thrilling creatures on the planet. No matter how well an animal is protected by nature and by evolution, it is always vulnerable to us humans. There's nothing we can't do when we put our minds to it. Still, at least we are now beginning to put our minds to saving the

11. GOLDEN POISON FROG

golden poison frog: we'd be much poorer without such a creature to give us nightmares.

The poison dart frogs, or to be more technical, the Dendrobatidae, are a bewildering bunch: bizarre, beautiful, deadly. Most could sit comfortably on your thumbnail, though such an arrangement might not be so comfortable for you. They come in a glorious range of colours and patterns because they need to be seen. Their lives depend on their ability to stand out from their surroundings. No point in being lethally unappetising unless you tell people: after all, if they discover it for themselves it's too late for you.* This is called aposematic colouration, and no creature adopts this tactic more stylishly. Each shining frog looks as if it has just been painted in Airfix paints and has yet to dry. It's not clear what makes them so poisonous, though it's probably something to do with diet – at any rate, they lose their toxicity in captivity. There are 175 species, and they're all toxic in different degrees. It is one of the great paradoxes that the wild world occasionally goes in for: let's not bother hiding. Let's tell the whole world we're here, let's shout our identity as loud as we can: and then we'll be safe. These fragile little specks of colour are so powerful only a bulldozer can hurt them.

---

* Natural selection works at the level of the individual, not of the species. There's no evolutionary point in doing something "for the good of the species". It's your own genes you need to propagate: not those of some conspecific stranger.

Les demoiselles du Waveney

I have a canoe. I keep it on the River Waveney, which flows with Norfolk on one side and Suffolk on the other. Not a kayak, a Canadian: a rather portly craft not entirely unlike Gerald Durrell's *Bootle-Bumtrinket*. You use a single-bladed paddle, preferably in a J-stroke to keep the damn thing straight. You don't do the Eskimo roll, at least not on purpose. And it's the finest possible way of getting close to the wildlife of the river. It's best in spring, when the warblers (seven species heard regularly) are in full song and the stinkboats mostly aren't. The river is navigable as far as the excellent Locks Inn; beyond that the canoes have it to themselves. After Bungay the river gets narrow and in places is closed over by a canopy. At others it flows, with the gentle courtesy the river rather specialises in, through broad floodplain water meadows with lush vegetation along the edges. And at times the river is alive with banded demoiselles.

They are pretty insects one at a time, but in hundreds they are devastating. Each of the four wings of the male is picked out with a blue-black inkblot, and it's a mark that adds great drama to these massed dances as the males compete for the attention of the females. You try hard not to think of it as the fairies' ballroom because that would be soppy: this is a great wet swarming sexy river, flowing gently and yet fizzing with the urgency of copulation. If they're fairies they're more like the sex-mad sinister creatures in *A Midsummer Night's Dream* than Tinkerbell. In a long, quiet paddle you can feel the earth teeming all about you: a great urgent need for each dancer to seize the opportunity of a lifetime: to dance in the sunlight and find a partner. It's a paddle to the very heart of life.

These are damselflies, related to dragonflies. They're all part of the order of Odonata, marked by long bodies, large multifaceted eyes and wings in quadruplicate: a double pair. They are formidable aerial predators, and are reckoned to be among the fastest insect fliers: there are claims for bursts close to 60 mph, 100 kph, though more usual reckonings suggest a flat-out speed of around 34 mph, 55 kph for the fastest species, and a cruising speed on 10 mph, 16 kph. They prey on mosquitoes, flies, bees and the occasional butterfly: anyone who has ever tried to catch a fly by hand must be impressed at the skills every dragonfly possesses. But they spend most of their lives as nymphs. Dragonfly larvae are as ferocious underwater predators as the adults are in the air. They use an extending jaw to snag their prey; the bigger species are perfectly capable of feeding on vertebrates like tadpoles and small fish.

Some dragonfly species are pretty big: the biggest of all is a damselfly, *Megaloprepus caerulatus*, found in South American forests, which can have a wingspan of 7.5 inches, 19 cm. But these are nothing compared to some of the extinct dragonfly relatives. The Meganeura of the Carboniferous era reached phenomenal sizes: a wingspan of 25.5 inches, 65 cm. It's as if the Carboniferous sky was a-fizz with Sopwith Camels and Tiger Moths. So why are there no insects as big now? Why don't the giant insects of horror films exist to terrify human communities? The usually accepted answer is oxygen. The insect breathing apparatus, the tracheal breathing system, soaks oxygen out of the atmosphere without need for lungs. That's wonderfully economical and efficient but it places a limit on overall size: oxygen can't travel too deep into an animal by this system. The Meganeura prospered because the Carboniferous world had a much higher percentage of oxygen than the modern world, which has a mere 20 per cent.

Adult dragonflies are defined as much by their eyes as by their crisp and efficient wings. The pair of compound eyes more or less cover the head: the animals seem to be all eye when viewed head on. Apart from bees, dragonflies have the best vision in the class of insects. Each eye has several thousand elements, in a system radically different from the one we vertebrates use. The system gives them spherical vision, and an ability to see beyond the human range in the ultra-violet wavelengths. It's suspected that dragonflies see the sky as a far more intense blue, the better to pick out flying prey, and for that matter, flying predators like hobbies, which are falcons with a taste for dragonflies.

The evolution of the eye has been a hot topic for scientists and creationists ever since Darwin published his big book, as we have already seen. The eye of the dragonfly is so clearly and profoundly different from a human eye that it poses a very obvious question: if your Great Designer designed an eye, why did he bother to design so many different kinds? Not a very economical process: not one of those parsimonious explanations that science and logic demand. The fact is that vision is so useful that the eye has evolved 50, maybe as many as 100 times over. There are many different ways of seeing the world: the dragonfly, with its compound eye, is a particularly vivid example of one of them.

The distant mountains
reflected in the eye
of the dragonfly

From the Japanese haiku poet Issa. The Bird's Nest Stadium at the heart of the Beijing Olympic Games – China, not Japan obviously – was always full of dragonflies during the Games. Not much wildlife seemed capable of making a living in that city, but there were dragonflies everywhere in the Olympic Complex: and as a result on two occasions I saw a hobby – a Eurasian hobby, not an oriental, since you ask – flying over the stadium. So that was another dragonfly epiphany. Dragonflies are good at that. They have always been popular with writers of haiku, so let's finish with one of Basho's.*

Red pepper
put wings on it
red dragonfly

---

* Translated by Patricia Donegan.

12. BANDED DEMOISELLE

A miraculous draught
of newts

I lived for a while in what might have been the most northerly house within the M25. It was built over a railway tunnel. The trains thundered below while the motorway thrummed in our sleep and cast an orange stole along the horizon at night. Barnet and Enfield were just down the road. Nevertheless, there was a sheep-field out the front and a small copse at the back. It was all right. And so naturally, we dug a pond in the garden. It was tiny: you could step over it. It was still a nice thing to have: blackbirds bathed in it and we planted flags and marsh marigolds. And one day there were newts. Just like that. There are three species of newts found in Britain – these were common newts. Their appearance was a small miracle. I knew of no other pond within walking distance: yet the newts undauntedly walked from it to reach my place. They travelled some considerable distance to seek out this delightful but tiny puddle of water. Who told them it was there? What drove them across the wastes of suburbia to find it? One day they weren't there: the following day the pond was jumping with them. They had made a long and extremely dangerous pilgrimage through the cat-thronged avenues of Hadley Wood, presumably sticking as far as possible to the dew-soaked grass, in order to find it.

It was then that I learned what it is to be an amphibian. They could leave the water at will, but only to find more water. Without moisture they don't exist, but on a cool, damp night they can be intrepid travellers, covering impressive distances on their short but efficient legs. I have been on a night newt hunt in a deeply surreal place outside Peterborough with the charity Froglife: an extensive area that was once mined for brick clay. The resulting pits filled up with water and then with crested newts, the rarest of Britain's newt trinity. The place has been put together without any logic that might help a human walker. It has a nightmarish sodium light from the main roads that surround it, and yet shine a torch onto the water and there beneath its surface I could see what seemed like a dance of the dragons: miniature submerged monsters that looked as if they might surface at any moment and start breathing tiny gouts of flame. I rather saw the point of newts that evening, even though an interest in newts is generally considered to be an admission of failure in life, the last word in wetness. Gussie Fink-Nottle is a disastrous character in PG Wodehouse. He is both a teetotaller and a newt-fancier: as if one were as bad as the other. "Gussie, a glutton for punishment, stared at himself in the mirror... many an experienced undertaker would have been deceived by his appearance and started embalming him on sight."

Newts and salamanders make up the order Caudata: a group of 580 or so species that look superficially lizard-like: low-slung, four-limbed and long-tailed. They live only in the northern hemisphere and the tropics. Most are pretty small, measuring 4–6 inches, 10–15 cm, but there are one or two biggies: the largest is the Chinese giant salamander which can reach 1.8 m, nearly 6 feet. Some are wholly aquatic, others make a living in moist earth. There are some picturesque individuals among them: a group called mudpuppies and olms, and another called hellbenders, which are found in the Ohio river system: there is a subspecies named the Ozark hellbender, fine name for a country band. The family of sirens has done away with hind limbs and retains only rudimentarily forelimbs. Salamanders have a rather more exciting tradition than newts in human culture. They are mentioned in both Aristotle and Harry Potter. The myth is that they are born out of fire, which always seems a sad thing to believe about a creature that loves cold, moist things. The belief is usually put down to the fact that salamanders like to lurk in hollow logs: put a log containing a salamander on the fire and it will come out pretty sharpish.

The axolotl, a creature that has won a certain amount of glory for the oddness of its name, comes from this group. It is a salamander found only in two lakes in Mexico; alas, one of these has been drained and the other reduced to a series of canals. The axolotl is now categorised as critically endangered on the IUCN* Red Data List. The threat increases as Mexico City continues its inexorable spread.

Caudatans have one particularly odd characteristic: they can regenerate entire limbs. The shedding and regrowth of the tail is a trait they share with lizards, but salamanders and newts can go one better: if they lose a leg they simply acquire a new one. This remarkable ability has fired the imaginations of scientists, who are experimenting with stem-cell technology to see if humans could manage the same trick. They are doing most of this work with axolotls. There is a sermon to be preached here: be careful which species you allow to go extinct, because you never know what use you will find for them. While there are better arguments against extinction than the fact that humans might one day find them useful, it's still a handy reminder of the value of diversity of life on earth.

---

* The International Union for Conservation of Nature and Natural Resources. The Red Data Book was established in 1963 and is accepted as the global authority on the status of species. Critically Endangered is the highest category before Extinct in the Wild.

Cannibal sex

Jean-Henri Fabre had a revolutionary idea. Instead of killing insects and sticking them on a card and trying to classify them, he watched them. For him, a living animal had more meaning than a dead one. He wrote about the lives of insects not with detachment but with passion. He was called "the Homer of the insects". Perhaps he was more of a Joyce in his innovative nature, the staggering originality of his vision and the unstemmable fountain of his prose: as if Molly Bloom were an entomologist. He never bought into Darwinism, though Darwin called him "an inimitable observer". It was observation that Fabre excelled at: that and an almost incontinent need to communicate his discoveries. He was a great loather of pomposity: what he cared about was living things and vibrant, meaning-filled communication. He was one of the great life-affirmers. He wrote: "Others again have reproached me with my style, which has not the solemnity, nay, better, the dryness of the schools. They fear lest a page that is read without fatigue should not always be the expression of the truth. Were I to take their word for it, we are profound only on condition of being obscure." I have endeavoured to follow this principle in everything I have ever written on wildlife.[*]

Fabre lived most of his life in Avignon, and his most vivid subject was the heat-loving insect-life of Provence. It was in

[*] Another line of Fabre I have taken to heart: "I have made it a rule of mine to adopt the method of ignorance... I read very little... I know nothing. So much the better: my queries will be all the freer, now in this direction, now in the opposite, according to the light obtained."

Provence that I first came face to face with a praying mantis. Quite literally so: I was crawling through the undergrowth in the Camargue with a view to getting close to a flock of flamingos. I was on a French exchange visit, those fraught occasions that are supposed to do one so much good. (It's a bit like an arranged marriage, I suppose, but without the sex.) I was brought up short by the mantis, which I showed to André, my French *ami*, in sudden wonder. Contemptuously – no doubt thinking I was exhibiting typically English fearfulness – he flicked at it with his finger and sent the damn thing spinning. I wasn't really frightened by it, though: a little alarmed, yes, but in a rather thrilling way. I hadn't thought that my eyes would ever alight on so exotic a creature: gazing back at me, apparently engaging in eye contact, with an oddly human expression: mild eyes, weak pointed chin and those "hypocritically" praying limbs, as Gerald Durrell wrote. In *My Family and Other Animals*, Durrell writes of a fight between Geronimo, a gecko, and Cicely, a praying mantis, that took place "above, on and in my bed". To come across one of the participants of that battle was a strange privilege.

The praying mantis, *Mantis religiosa*, is one of 2,400 species of mantis. They all possess the strange humanlike face and the savage grasping arms, crooked in apparent prayer: raptorial legs that can reach out and clasp prey in an instant. Almost all mantises work as daytime ambush predators, using exceptional eyesight. The queer arrangement – an eye on each of the upper corners of the triangular face – gives a large area of stereoscopic vision: that is to say, two-eyed vision with real depth perception. The characteristic bobbing and swaying motion of the mantis is reminiscent of the way a kingfisher constantly ducks its head. This shifting of angle is used for the same reason: to get a better three-dimensional fix before striking. An ambush predator needs to be hidden, and mantises go in for camouflage, most of them adopting an all-purpose green or brown, but some tropical species operate elaborate disguises to become leaves and flowers. This tactic helps them both to prey and to avoid predation: the luxury of the alpha predator is unknown to those lower down in the Greek alphabet. A lion doesn't have to look for a safe place because a lion *is* a safe place. Not so for a fat and juicy mantis.

The edibility of the mantis is perhaps the most famous thing about them. They go in for sexual cannibalism. Fabre observed a male mantis actually offering his head to a female. If she accepts, it means that she can snack on his brain while he copulates. A male can copulate long after his brain has been devoured. The female carries on eating while the male carries on copulating, eventually, as Fabre said, the two "become one flesh in a far more intimate fashion". He had one female under observation who devoured seven husbands in a week.

Subsequent studies have demonstrated that a male, when deprived of his head, becomes a much more vigorous lover. He is capable of copulating twice as long and his chances of fertilising the female are doubled. The best chance of getting your genes to live on, the best chance of becoming an ancestor, is to leap on top of a female and lose your head in the process. There is a clear opportunity here to make all kinds of analogies with human life, but I shall leave that job, dear reader, to you.

Beautiful shirts

And there are still more of them. More things of beauty and wonder, an eternal brilliant bewildering rain of them. Like Gatsby's shirts: "He took out a pile of shirts and began throwing them, one by one, before us, shirts of sheer linen and thick silk and fine flannel, which lost their folds as they fell and covered the table in many-coloured disarray. While we admired he brought more and the soft rich heap mounted higher – shirts with stripes and scrolls and plaids in coral and apple-green and lavender and faint orange, with monograms of Indian blue. Suddenly, with a strained sound, Daisy bent her head into the shirts and began to cry stormily.

"'They're such beautiful shirts,' she sobbed her voice muffled into the thick folds. 'It makes me sad because I've never seen such – such beautiful shirts before.'"

A famous scene in *The Great Gatsby*. As I write this book, there are times when I feel like Gatsby, revelling in showing off such richness; at other times more like Daisy, overwhelmed by what I am looking at. Many novelists, myself included, have experienced that wonderful feeling when a story and a character take on a life of their own, apparently quite independent of the author: the characters speak their lines from the depths of their own personalities and the story unwinds in a certain direction because there is no other way in which it can go. The novelist seems, during these suspended, fragile passages, not to be writing as much as taking dictation. Writing this book has sometimes been like that, only more so, for none of it comes from my imagination, conscious or otherwise. Life's endless forms just come welling up from the earth as Gatsby's shirts rained down from the sky. All I have to do is try and record some of them as they describe their brief parabola before us.

Which brings me to the caecilians. They're amphibians, but not as we know them: nothing like frogs or salamanders or newts. They are the amphibians of obscurity, creatures who mostly live beneath the earth, seldom seen, not understood overly well, not attracting much interest when they turn up. They don't look like frogs and salamanders: they don't look much like vertebrates at all, more like earthworms. They even have a segmented appearance, because they carry ring-shaped folds in their skin. Some are as gorgeous in colour as Gatsby's shirts: blue, pink, orange, yellow, striped like a humbug. No monograms, though. More are drab earth colours. They are not beautiful in human eyes, not like the blue morpho butterfly, but all the animals in this book, just like all the animals not in this book, have something of wonder about them. The secret amphibians are as wonderful as birds of paradise and tigers.

There are around 200 of them in nine families, though naturally, there are rival methods of classifying them. Most caecilians are burrowers: talented and effective, with a strong skull and a pointed snout. They have no limbs, and their eyes can do no more than tell them whether nor not it's dark: important information because a caecilian that came up to the surface in daylight would be extremely vulnerable. They are found in the tropics around the world in appropriately wet places. Their strong suit is smell: they have retractable tentacles between the eyes and nostrils that collect and transmit chemical signals. They range in size from 12 cm, just over 4 inches, to 1.6 m, more than 5 feet. The burrowers emerge at night and hunt for earthworms, termites and other inverts. Most of them fertilise their eggs inside the body and the female gives birth either to live larvae, or in some cases, miniature adults.

Thus a large group of terrestrial vertebrates – creatures of our own kind – manages to make its living without impinging at all on human consciousness. There are no caecilian stories, no caecilian myths, no caecilian recipes. They are creatures we haven't even noticed: and could lose entirely without noticing. Question: would we be the poorer for their loss? It's like the sound of the tree that falls in the deserted forest: does an extinction matter if humans were scarcely aware that the creature existed in the first place? That's one for philosophers. Here's my contribution to the debate:

It's not all about humanity.

Unreal city

So there's yet another form on the internet you have to fill in. Address, it demands, starring the space as a compulsory field. Insisting on a house number, even if your house hasn't got one. And then there's a space labelled "city". Because we all live in cities, don't we? That's at least slightly true. The assumption that a person lives in a city stands up pretty well in a rough and ready sort of way. City has become the default mode of living. The modern city is where humans get on with the business of life: Shanghai, 18 million, Istanbul, 14 million, Karachi, 13 million, Moscow, Mumbai, Tehran, Beijing, all with 12 million. New York and London have eight million each: rural hamlets, merely. How long have we been living like this? Depends on how you define city: at the first establishment at some river crossing? At the first million? At the first high-rise building? Two thousand years, 200 years, since the end of the World War Two? In evolutionary terms, any one of those periods is an eyeblink. We evolved as savannah-dwellers, to live in extended family groups and small gatherings called tribes: now we're all on top of each other, living among strangers and wondering why life is sometimes so difficult and so out of joint.

Termites have been at it for millions of years, probably since the early Jurassic, which is to say around 150 million years ago. They do the large community thing in a way that dismays human sensibilities. The lives of individuals are nothing in a termite colony: it is all about the good of the community as a whole; the continuation and spread of the genes represented by that colony, to put the thing in evolutionary terms. Self-sacrifice and suicide are routine aspects of life in many termite colonies. The obvious and inevitable comparison with human societies is deeply disconcerting.

Termites build on a massive scale. The wooded savannahs of Zambia are studded with termite mounds: they are as much a feature of the landscape as the trees. My old friend Phil Berry has calculated that one per cent of all Zambia is a termite mound. I have frequently climbed to the top of one to get a better idea of what was happening all around me: any small eminence on a floodplain is a useful thing. I have seen lions doing the same thing, for they love a good view. The big old male baboons often stretch out on a termite mound to keep watch as the rest of the troop forages. Termites have done as much to create this landscape as the tree-clearing elephants.

And by recycling organic matter into their edible selves they play a crucial part in the ecosystem. Termites feed on dry and damp wood, on leaf litter, on dung and on grass. There are around 2,600 species of them, all in the hotter places of the world. For termites, it's all about the nest or mound or city: they define themselves by their habitation; constructions that can be so vast that it is impossible to believe it was the labour of so small a creature. Great domes rise 10 feet, 3 m, above the savannah, often with a vast and centuries-old tree growing from them. In some places termites build baroque sky-reaching structures up to 9 m or 30 feet tall, looking a little like the Sagrada Familia cathedral in Barcelona. In Australia compass termites build blade-shaped nests on a north-south axis: they catch the sun for its warmth in the morning and again in the evening but minimise its worst during the heat of the day.

All termite structures are built around a complex system of tunnels and chambers and shafts. Convection currents function as air-conditioning. The airflow keeps the temperature in the brood chambers to within a single degree of the optimum. Humidity is maintained even in the dry season by controlled condensation. Some of the more ambitious species go in for underground gardening, farming a fungus that has evolved along with the termites. It is nourished by the excrement of the inmates and is self-perpetuating: the spores from the fungus germinate on fresh faecal pellets. How did this deeply complex system begin? No doubt it was some accident or series of accidents, one that worked for mutual profit and so became, as it were, formalised into the lives of the termites. All it takes is a few tens of millions of years and you have a system of uncanny perfection.

Termites are eusocial animals, like the two species of mole rats we've already encountered: that is to say they work on a reproductive division of labour – not all members of the colony are involved in reproduction. There are overlapping generations and they go in for cooperative care of the young. Many species of wasps, bees and ants are also eusocial animals.

Termites are divided into castes, which will include an egg-laying queen that can live for 45 years, and sometimes a king, who will service her all her life. There are workers, and some species have soldiers. In some cases, these soldiers have armour and pincers so large and ferocious that they can't feed themselves, but must, like juveniles, be fed by the workers. Their main job is to defend the nest from marauding ants: when there is an invasion, one soldier will block the breach until he is killed, and his place is taken by another. When the gap is too great for a single hero, the soldiers form a wedge or phalanx and make a wall to keep out the invaders, while the workers repair the damage behind them. In plain terms, they go into battle to die. One species, the tar baby termite, repels invaders by means of suicide. The soldiers lethally rupture a gland in their own bodies to release a yellow fluid that engulfs the invaders.

Termites also produce sexual forms with wings, creatures that set off in search of adventure and to form a new colony.

In the Luangwa Valley they will do this when the rains come at the end of the six-month dry season, producing a bonanza of flying insects: so massive a spewing-forth of protein that martial eagles, one of the most majestic birds in creation, will go slumming for termites and sit by the exit of a mound to gorge themselves on these would-be sexual adventurers. Many die: but new colonies are still formed and the savannah still undulates with termite mounds.

It is a perfect society: the totalitarian dream in action. The only flaw in it, so far as human society is concerned, is that humans are not termites. Perhaps 150 million years of city life will bring about some changes.

"Fish"

You know the way penguins leave the sea when there's a leopard seal in the water? Whizzing up headfirst and somehow landing flat-footed on the ice cliff above? Popping up together like a thousand champagne corks? Looks impossible, I know, but it's not actually all that hard. I've done it. More or less, anyway. It happened when I was snorkelling off Bali. We set off as the sun rose in a small boat propelled by a charming Balinese (more or less a tautology). At this length I can't remember his name, but it was probably Ketut: there are only four names in common use on that lovely island. So when we got to the coral I lowered myself off the boat and spent a happy hour or so among the inhabitants of the reef, taking in the gloriously biodiverse colours of the community: as if all the richness of the Animal Kingdom were laid before me. I cruised weightless above it all. At times I folded in half like a clasp-knife to dive, getting splendidly deep with just a few flips of the fins, negatively buoyant as I am. Allowing myself to drift back to the surface, I cleared the tube with a great Moby Dick blast and then trudged on, passing from one wonder to the next.

And then in the space of a second I went from trudge to boat. Bang. There was scarcely any time-lapse at all between the sighting and the boating: no interim period of decision-making, swimming back to the boat and pulling myself out. It was an instantaneous transition. Shark–boat. And there was Ketut laughing amiably: "No eat!" he said. "No eat!"

"It's all very well you sitting in the bloody boat and saying no bloody eat," I said. "No. Home. Breakfast." And still laughing at me, though ever so kindly, Ketut took me back to the shore.

Leaving me wondering forever after about my reaction to that shark. I can see it still, with immense vividness: one of those unfading snapshots, like the Definite Male Lion. But unlike the lion, this shark was, as Ketut said so eloquently, non-lethal. Not to humans, at any rate. It was a healthy size, a good 6 feet, but it was never likely to take on anything of a similar length. It swam directly beneath me, going about its own business with immense driving purpose. There was nothing to be afraid of. Why, then, was I afraid?

I still don't know why it affected me so powerfully. Perhaps it was because of all the received information about sharks: the way we have turned them into fabulous monsters, film stars, bringers of death, far more fearsome than the far more lethal mosquitoes or for that matter, jellyfish. Perhaps fear of sharks is a purely modern phenomenon.* But perhaps not. Perhaps there is something atavistic in humans that responds to the sight of a shark. But there is a third possibility. Perhaps it was not the shark *itself* that inspired such decisive evasive action. Perhaps it was the fact that it was so completely different from every other living thing around me. It broke the pattern of normality: and that can trigger the flight-response in a prey animal. Many a time I have sat on a horse that has spooked and even tried to run from a wheely-bin: not because wheely-bins are terrifying but because this one wasn't there yesterday. Wheely-bins are not threatening: unfamiliarity is. If it breaks the pattern, it's time for action: every horse believes in the principle of better fast than sorry. In short, was it my over-civilised self that responded to the culture of *Jaws* and got me out of the water? Or was my ancient ancestral self seized with an ancient ancestral fear?

The shark was spectacularly unlike any other fish I had

---

* In 2012 there were 118 shark attacks reported, of which 17 were fatal. Malaria kills 2,000 to 3,000 people every day.

set my eyes on that morning, and there had been plenty. It moved in a different way. Most fish move with a rapid, almost imperceptible wriggle: this shark moved in long, bold, side-to-side lashes. Everything about it proclaimed its difference. The cut of its jib, what birders call its jizz, its mien, the impression it gave: all was different. This was not a fish as other fish are. It was another class of being entirely. And so in that one lightning moment, Zen-like, I was enlightened. Not about the meaning of life, but certainly about the meaning of fish. I understood why there is no such thing as a fish. In other words, after about 100 chapters and nearly 100,000 words, we have got the punchline of the fish joke. Why is there no such thing as a fish?

The fact is that "fish" is not a coherent term when it comes to taxonomy. It used to be thought that there were five classes of vertebrates: mammals, birds, reptiles, amphibians and fish. Now it's all been changed. For a start, many classify reptiles and birds together, but what matters here is that there is no single class of backboned gill-breathing swimming animals. In fact, there are four of them. Each has found an approximately similar solution to the problems posed by an aquatic life: the same streamlined shape, a similar side-to-side swimming motion, the same way of gathering oxygen from the water. But these four classes are no more closely related to each other than we are to the red-eyed frog or to the kori bustard.

Four distinct classes make up the group formerly known as fish. The biggest one is the ray-finned fish, which include goldfish and are what we normally think of when we hear someone say "fish". The second are the cartilaginous fish, which include sharks and rays. And there are two more obscure classes with far fewer members: the jawless fish and the lobe-finned fish.

A fishmonger is of course a person who mongs fish. The nonsense in that last sentence is not the penultimate word, but the last.

True bugs suck

It's the sound of the heat itself. The sound that proclaims an elemental truth: that no one should be expected to do anything but lie in the shade until it's over. That's the cicada, celebrated by Gerald Durrell in *My Family and Other Animals*: the insane heat of a summer afternoon on the island of Corfu rendered into sound by the indefatigable insect. It is a sound that thrills the traveller from a colder place, at least at first. Later it perhaps exasperates, soothing as fingernails on a blackboard, but finally it becomes the essential voice of the place. Cicadas turn up all over the world in the warmer places, and they love to sing, especially in the heat of the day when everything else is silenced.

> Stillness —
> the cicada's cry
> drills into the rocks

Another haiku from Basho: an insect epiphany. A cicada is a *semi* in Japanese but there's nothing half-hearted about a cicada. Some of the species are pretty hefty: on Lamma Island they always seemed to be crashing into me, like flying Volkswagens, interrupting early-evening drinks with mad forays into human company, headbutting people, bashing into drinks and lying stunned and enormously winged on the table before they lumbered back into the air and flew off to continue their eternal chorus.

Perhaps cicadas bug you. If so, you have the blessing of science this time: cicadas are members of the order of

Hemiptera, otherwise known as true bugs. They are named for the hard base to the forewing; Hemiptera translates as "half-wing", but that's not so terribly interesting. What true bugs can do, probably better than any creature that has ever evolved, is suck. They have sheathed mouth parts which form a beak – technically a rostrum – that can pierce tissue. Once the piercing has been achieved, they suck. Mostly they pierce plant tissue and suck sap. Aphids, which include greenfly, insects many gardeners wage chemical warfare on, are bugs. So too are the insects that cause cottony cushion mould, which has devastated citrus crops.

Some bugs prefer to penetrate animal tissue: bedbugs are rightly named, being true bugs. There are other bugs that suck vertebrate blood, some of which are vectors for Chagas disease, so it is possible that it was a South American bug that ruined Darwin's health. The most picturesque members of this group are the assassin bugs. They prey mostly on their fellow invertebrates, and they can take on creatures much bigger than themselves. They do so by what is called external digestion.* They use the rostrum to penetrate their victim and inject a saliva that pre-digests its insides. They then suck it up like a child with a milkshake.

There are more than 50,000 described species of true bugs, some say as many as 80,000. That's because the classification gets a little complex and controversial. But we can take a moment to admire the versatility of the Hemiptera, for the bugs have come up with many ways of facing the world. They include the water boatmen and water scorpions, and one of the very few marine insects, the Halobates or sea skaters.

Then there is the 17-year locust, which is not a locust† at all, but a cicada. There is also a 13-year locust, another cicada; both of these misnamed species are found in the

---

* Spiders, as we have seen, use a similar technique.
† True locusts are related to grasshoppers.

United States, and as their names suggest, they appear once every 13 or 17 years. Like a plague, in fact, hence the name, given by people who knew the Bible better than the works of Linnaeus. These creatures spend the first 13 or 17 years of their life underground where, in the manner of bugs, they live by sucking, penetrating tree roots to suck out the sap. But every 13 or 17 years there is a sudden emergence: a dramatic explosion of rock-drilling sound. It's celebrated in the Bob Dylan song "Day of the Locusts". The great screaming dance of the cicadas continues for a couple of weeks, and naturally, it's all about mating and egg-laying… and then the frenzy is over for another 13 or 17 years.

Why pick such a number? Stephen Jay Gould, already met in these pages, says: "Many potential predators have two- or five- year lifecycles. Such cycles are not set by the availability of cicadas (for they peak too often in years of non-emergence), but cicadas might be eagerly harvested when the cycles coincide. Consider a predator with a life-cycle of five years: if cicadas emerged every 15 years, each bloom would be hit by the predator. By cycling at a large prime number, cicadas minimize the number of coincidences (every 5×17, or 85 years, in this case). Thirteen- and 17-year cycles cannot be tracked by any smaller number."

This argument has been disputed, but it has a certain beauty about it, so I rather want it to be true. It has precisely the kind of mathematic perfection that is always so enticingly just out of reach to a non-mathematician. All the more so because in normal circumstances, numbers rather bug me.

The stillness of salmon

A great sigh of disappointment at the nearest of near misses. You can't not utter it: it's part of human nature, part of animal nature; it seems torn from your very guts. And then another one – ahhhh, no! So close! But then comes a third, breaking the surface like a rocket, like a penguin only better, firing itself at the sky, wriggling like a mad thing, contorting itself in the air like a long-jumper seeking those extra centimetres – perhaps as many as 20 – with a hitch-kick technique, touching down and getting half an instant of purchase in the foaming exuberance of the summit – and for a moment it seems it's going to be hooshed back to the bottom, but no, after a series of glorious triumphant shuffles in which its mouth seems to make contact with its tail on both sides of its body in less than a second, it's through, cutting though the opposing waters of the pell-mell stream like an agile blade.

That is a salmon taking on all the water in the world as it scythes and knives its way upstream to the spawning ground, driven from the sea by the irresistible call to make more salmon,[*] a sight I witnessed at the Falls of Shin in Scotland. A salmon seems at first sight to be the perfect representative of the ray-finned fish: the great voyager, the unstoppable swimmer, the fish that's crossed the barrier between salt water and fresh, and what's more, is considered supremely edible. Salmon are also supremely fast and skilful, but at

---

[*] Bill Bryson, in one of his travel books, draws a cartoon of two salmon at the foot of the waterfall. One says to the other: "Sod it, I think I'll just stay here and have a wank."

the end, they are supremely helpless as, the spawning done, they have nothing more to do with their lives except die. I have seen them hunted down in British Columbia by bears, and it's a harder job than it looks while they're still on the move. One of the bears I saw was presumably young, and certainly spectacularly incompetent: floundering after the fish like a Labrador, all misdirected enthusiasm and playful bounces. Eventually it settled down to look for pre-killed, half-eaten salmon that had been caught by bears who had mastered the art, eaten the tastiest bit, left the rest and moved on.

The image of the salmon leaping the fall is the most romantic in the fishy universe, a sight that spells out the truth that when it comes to motion through water, nothing is quite as good as a ray-finned fish. To tell someone that he swims like a fish is the most reckless compliment. Nothing has mastered the water quite as much as a ray-fin – and yet motion is not their masterpiece. The ray-finned fish's true mastery of the water comes from stillness. In the motion-less goldfish in the aquarium, suspended between surface and gravelly floor, unmoving, contemplating life in the profound way that only a goldfish can: that is where you find the true greatness of the ray-finned fish. Their secret is the swim bladder: with it, a fish can retain neutral buoy-ancy at any depth, not rising, not sinking and not needing to move a muscle. It can lean on the water as you lean on a lamp-post, it can lie in the water as you lie in your bed, and yet it can shoot off through the water as Usain Bolt shoots off along the track. More than to any other creature on the planet, water is home to the ray-fins.

The swim bladder is part of us, too. That was one of the insights of Charles Darwin: "There is no reason to doubt the swim bladder has been converted into lungs [and that] all vertebrates with true lungs are descended from an ancient prototype." Every breath we take reminds us that our ances-tors are watching us, telling us of our line's piscine past.

Most vertebrates are ray-finned fish. There are almost 30,000 species described. With a few odd exceptions – thermal pools, the Dead Sea – wherever there's water there are ray-finned fish. One of the weirdest and biggest is the ocean sunfish, a creature which inexplicably likes to bask flat-out on the surface of the sea. I saw one from a boat off the coast of Cornwall, lying there like an enormous flabby dinner-plate. They can reach 11 feet, 3.4 m, across and weigh as much as half a ton, 1,000 kg. There are plenty of contenders for smallest; anyone who has visited an aquarium has marvelled at the little shards of flickering subaqueous light that are fully-formed adult fish. The prize perhaps goes to the dwarf pygmy goby at 0.3 inches, 7.6 mm.

The weird thing about the great biodiversity of ray-finned fish is that they all look pretty much the same. That's inevitable, because travelling through water is a hard job, much harder than moving through air. That makes streamlining essential, and since there is really only one way of streamlining there is no vast divergence in body plan. The class of mammals brings us giraffes and elephants and hippos and mice and bats and kangaroos. The ray-finned fish bring us salmon and perch and carp and sturgeon and pike and catfish and cod and mullet and they're all roughly the same shape. That rather blinds us to the great diversity of fish. It's also true that, not being aquatic animals ourselves, we are never as intimate with fish as we are with birds and mammals. The ray-finned fishes are the great undiscovered ocean of biodiversity among us vertebrates.

Let copulation thrive

The simple fact of flight: is there anything to touch it for sheer glory in the wild world? Birds, more than any other group in the Animal Kingdom, have fascinated humans across the ages because they can fly. Among all the inverts, it's butterflies that enthral humans above all others because they are beautiful and they fly most conspicuously in the daylight. When we talk of the human spirit taking wing, we mean nothing but greatness and glory and optimism. James Joyce called himself Stephen Dedalus in *Ulysses* and in *A Portrait of the Artist as a Young Man*: naming himself for Daedalus, the man who invented human flight, whose son Icarus flew too close to the sun.

How many times, I wonder, have I raised my eyes to the heavens to gaze at some avian superstar: say, a bateleur eagle, gliding in glory across the valley, as I pause, just for a moment, to slap irritably at some bothersome insect. How irritating is a bloody fly when my soul is given up to the glories of flight? And yet nothing flies like a fly. Observe the houseflies in your kitchen: their complete mastery of three-dimensional space. They can be fast and direct; they can swerve and jink at speed, as anyone who has tried to kill one well knows. They step into the air as you and I step onto an escalator: a transition so straightforward it's hardly a transition at all. They can alight on anything, floor, wall, ceiling: it's all one to them; the transition back down to solid surfaces is just as simple. They are swift, agile and wholly controlled: surely no other group of animals on the planet is so completely at home in the air.

comfortably off humans who can afford to be finicky, and we'd soon stop that nonsense if we were to experience real hunger.

The beauty and glory and wonder of flies tend to pass us by. But that's true of so many species. There are ten million and more out there: how many of them do we really know, how many of them do we take delight in? And yet, far from any widespread human understanding, or even awareness, there are creatures as wonderful as any of the superstars, living out their lives without need of human endorsement. The fly on my window is just as brilliant – in some ways just as beautiful – as the nesting swan I can see on the other side of the glass as I write these words.

The Eden fish

Nature is benign. How could anybody deny it? All you have to do is find a nice crowded bit of coral and go looking for cleaner fish. It is so striking an example of kindness, sharing and generosity that you feel infinitely capable of taking the next logical step and believing in fairies. It is the favourite fish of the creationists, who say that so wonderful a system simply couldn't have come about by a natural process – or "by accident" as they prefer to say. The whole set-up is so striking that there are moments when even a grown-up person could almost believe them.

Cleaner fish make their living by tending other fish. A fish, often a much larger animal altogether, will submit to the attentions of a cleaner fish, lying still while the fish grazes over its body, nibbling away dead skin and external parasites. The cleaner fish gets a good meal, the big fish gets cleaned up and swims away happily. It does so without helping itself to its helper as a post-spa snack. Cleaning behaviour has evolved separately in a number of unrelated groups: wrasse, cichlids, catfish and gobies; there are also some shrimps who have taken on the job. Most of them have come to resemble each other: blue with a pronounced longitudinal stripe and an attenuated shape. Many will perform a small dance to advertise their services. A large predatory fish, the sort that would make a meal of a fish the size of a cleaner without a second's thought, will rest motionless while the cleaner does its work: a celebration of mutual trust as well as mutual advantage. The cleaner will touch and soothe a big fish before and during the cleaning process: a predatory fish will be touched about three times more frequently than less dangerous fish.

The presence of a cleaning station makes the surrounding waters a safe haven for other species. The suspension of normal life in such a place makes it ideal for the sudden entrance of a major predator, but that doesn't happen. Here is a corner of the wild world, a corner of fierce and unforgiving nature, a place that is supposed to be red in tooth and fin, giving itself over to the pursuit of peace, to everybody's mutual advantage. This is the myth of Eden acted out for us beneath the waves like an octopus's garden in the shade.* The idea of Eden goes very deep in most cultures: the place where the wild world leaves off its ferocity and the lion lies down with the lamb† and humans can wander naked without fear or shame.

The beguiling and complex life of the cleaner fish tells us that all this can be so: that peace can break out at any moment, just as Hathi the elephant called the truce during the drought in *The Second Jungle Book*, in the story "How Fear Came". Under the truce, predators and prey animals agree to share the riverbank to drink and rest without eating each other or running away, so that life for all might continue until the rains come again at last.

It's a wonderful story and like all wonderful stories, it contains a great human truth that goes very deep. So let me introduce the sabre-toothed blenny. This is a fish that looks exactly like a cleaner fish and even does the same advertising dance as a cleaner fish. But when the bigger fish stretches itself out to be groomed, the sabre-toothed blenny darts in and grabs a mouthful of healthy fish and shoots off

---

* There is no such thing as perfection in nature, nor should there be. Every Beatles album has its Ringo track: and is the richer for it.
† The beatific image of life's peaceful perfection is as vulnerable as a fish that has given itself over to the attention of a cleaner. DH Lawrence said: "No absolute is going to make the lion lie down with the lamb unless the lamb is inside," while Woody Allen said: "The lion shall lie down with the lamb, but the lamb won't get much sleep."

at high speed – thereby creating as big a problem for moralists and creationists as it does for the big fish.

So let me briefly assume the role of moralist. The first moral is that there is no moral. The second moral is that anyone who seeks moral lessons for human life in the wild world needs to be somewhat selective, or better still, blind. The third moral I will leave to James Bond. Here is an extract from the fifth chapter of *Goldfinger*: "A second reason why Bond enjoyed the long vacuum of night duty was that it gave him time to get on with a project he had been toying with for more than a year – a handbook of all secret methods of unarmed combat. It was to be called *Stay Alive!*" And that, in two words, is the only real moral lesson taught by the wild world. Stay alive, so that you might breed and become an ancestor. Only once you have done so is it suitable to die like salmon in the upper reaches of the rivers they were spawned in, their biological destiny fulfilled.

As humans we need a moral code to live by. We will spend the rest of our existence as a species arguing what that code should be and where our moral duties lie – but we won't argue nearly so much about whether or not a moral code is necessary. Unless you are a philosopher or a sociopath, you accept without needing to be told that we, as individuals and as a society, need a moral code.[*]

The wild world does not tell us that humans can, if they wish, abandon all moral codes, nor does the wild world teach us what that moral code should be. Human wives do not eat the heads of their husbands as a matter of course; nor is there any compelling reason for humans to model their society on a termites' nest, on an underwater cleaning station, or for that matter, on the sabre-toothed blenny. In

---

[*] Ivan in *The Brothers Karamazov* states that if you don't believe in God, all is permitted. But in an increasingly non-religious society, the concept of morality is as strong as ever, even if moral standards have changed. As a sportswriter, my job is to write about morality as much as the decline of the 4–4–2 formation.

this book I have constantly stressed the human continuity with our fellow species: and if there is a moral to be drawn from that continuity, it is to do with respect, compassion and generosity, and even love, not untinged by a certain self-interest. But this continuity does not require us to use selective animal behaviours as handy moral fables to stress certain ideas of what humans should be and how we should live our lives. That simply doesn't work.

13. ANOTHER HUMAN

Prostitutes and clients

Monty wears a radish in his button-hole, grows vegetables in exquisite pots in his exquisite flat and happens to think the cauliflower more beautiful than the rose. He's the mad uncle in *Withnail and I*, a film full of quotes that can be hurled like javelins. Withnail (mendaciously) tells Monty that he and his friend grow geraniums. "You little traitors. I think the carrot infinitely more fascinating than the geranium. The carrot has mystery. Flowers are essentially tarts. Prostitutes for the bees..."

The point could hardly be made better. Tarting is a flower's job, just as being a flower's client is a bee's job. Joyce nailed this relationship for all time in a single word: fleurting. Bees and flowers and flowers and bees: they've become more or less inextricable, the evolution and spread of one dependent on that of the other. Flowering plants existed before bees, but bees abandoned the traditional carnivorous diet of the order Hymenoptera and took on flowers as a specialised job: accepting lavish bribes of nectar and making off with weighty loads of pollen to feed the hordes back at the hive – and at the same time performing the essential cross-fertilisation on which flowering plants depend. Sex: we'd be extinct without it. Perhaps more than any other group of animals on the planet, bees make the world we live in. They make it possible, and they make possible its continuation. That's as true for humans as it is for anything else that lives on land, as we have seen with the bank-busting estimates of the commercial value of the pollination services of wild animals. (Here's a couple more: the value of UK pollination services is £510 million a year; to do the job of commercial pollination without them would cost £1.8 billion a year.) When it comes to pollination, bees dominate the market. There are about 20,000 species of bees, and they belong in the same order as wasps and ants. Their basic flower-foraging lifestyle is well known and found again and again, but there are plenty of exceptions; vulture bees specialise in carrion and are a rare example of non-vegetarian bees.

Bees brought me close to murder on one occasion and on another, to an encounter with potentially lethal consequences. Both occasions obviously involved Bob, my companion of the Definite Male Lion episode. In the first of these I was doing the cooking at our camp in the Northwestern province of Zambia, where we were looking for a Small Brown Bird. I was frying potatoes in an air temperature fractionally lower than that of the oil. It was a bad idea. A cloud of large bees made a threatening and bothersome circle around my head, and a dense swarm of mopane bees – little black bastards the size of pinheads that gather round your eyes and nostrils to sup your liquids – adhered to my face. Bob then made the mistake of uttering a critical remark about my cooking method. I suggested that he went and looked for the Small Brown Bird, now, and on his own. Either that or face a faceful of boiling oil. On another occasion, investigating a hollow baobab, we were set on by the famously deadly African bees. I was stung twice as we sprinted to the vehicle. The experience itself was not all that deadly, I have to confess, since it's generally accepted that a minimum of 500 stings are required to kill a human, but it was jolly painful and of course entirely Bob's fault.

The renowned "killer bees" of the Americas are an experimental hybrid of African and European honeybees. They got loose from the lab and went feral; they are more soberly referred to as Africanised bees. They are no more venomous than European bees, and for that matter, nor are the Africans who caused me such discomfort. But both African and Africanised bees are more inclined to defensive measures when disturbed: they will send out four times as many defenders and pursue for much greater distances.*

An African bird, the greater honeyguide, will perch in front of a human strolling the bush and utter a very particular prolonged chattering call. I've had the song sung to me many times, but have never followed it up. It is an invitation to follow. If you do, he will lead you to a bees' nest. The deal is that you can feast on honey while the bird gorges on the larvae and also on the wax, being one of the few animals capable of digesting the stuff. The story is that if you take the lot and leave nothing for the honeyguide, the next time it sees you it will lead you to a lion.

Many bees form eusocial colonies, though there are plenty of exceptions. Some bees take on arcane trades; there are mason bees, solitary bees that construct nests from clay, and carpenter bees, which bore into wood to make nests. The carpenters are not entirely solitary: often a female will live with daughters or sisters in a rough and ready social system.

The extremely organised nature of the eusocial bees has led to some extraordinary developments, and the most remarkable is the famous waggle dance (so famous that the brewers Wells named one of their beers Waggle Dance simply because it is produced from honey). The dance is about communication: successful foragers convey information to the rest of the hive about the direction and distance of a source of nectar and/or pollen, or water, and also on occasions the possible location for a new hive. This dance poses deep questions about the definition of language and the nature of human uniqueness.

---

* The greatest bee story in fiction is "Red Dog" from *The Second Jungle Book*, which tells how Mowgli, who likes to "pull the whiskers of death", uses the wrath of the bees to defeat a vast troop of invaders.

The more urgent questions to do with bees relate to their catastrophic decline in recent years. There are many causes: in the United States a combination of a virus and a fungal infection has proved fatal. There are problems connected with the use of pesticides and with the decline of flower-rich habitats. Two species of bumblebee have gone extinct in Britain over the past 75 years. In Britain, two-thirds of all pollinating species are in decline and 250 of them are in danger of extinction. Britain now has a Bumblebee Conservation Trust:* an organisation devoted to looking after creatures we once took for granted. And that is at the heart of the problem: that as humans, we still think with minds that evolved on the savannah. We *know* that bees and flowers have always been here and always will be here. Our brains can't deal with any other possibility: it is too absurd. The fact is that if we lose bees we not only lose the world we know, we also lose the world that supports our existence.

---

* I'm honorary vice-president.

That's no parasite: that's my husband

There is nothing that seems quite so much like the life of another planet – another solar system – another galaxy – than the deeply deep depths of the oceans. For humans anyway. For some creatures these places are as homely as the African savannahs, or to update the notion of the human comfort zone, Welwyn Garden City. But the deepest depths boggle the human imagination: the limitless three-dimensional environment of black emptiness, where most light fails to penetrate, where the pressure of the water would crush a human like a nut, where the only external nutrients come down in a gentle rain from the sunlit levels. Here dwell some of the strangest creatures on the planet: the most alien of all the aliens and in the most alien environment.

The natural human response to this environment is baffled rejection: how could any creature bear to live there? How could any of us vertebrates contemplate the fantastic possibilities of this place? How could we dream of facing the vast problems of life in so lifeless an environment? But these are human questions: they don't make sense for anything except human intuition. The fact is that evolution doesn't always seek some human idea of the good life: life exists because there is an opportunity for something to live. Stay alive! When such an opportunity can be found, you will generally find that some creature has evolved to take it up: often doing so in ways that scramble the human mind and confound the human imagination. We humans are creatures of light and warmth: naturally, we fail to understand those for whom cold and darkness has the same life-giving properties. What we see as a world of endless black horror is to the anglerfish the very breath of life. I am reminded again of the gnomes in *The Silver Chair*, who constantly recite the formula: "Many sink down to the underworld: and few return to the sunlit lands." The underworld is a fantasy from the mind of CS Lewis: but it actually exists at the bottom of the sea: in the aphotic zone where less than one per cent of the sun's light penetrates. Here you find the giant squid, already met in these pages, and also the anglerfish.

Anglerfish comprise 16 to 18 families. They are characterised by a lure that hangs over a wide mouth, not unlike the lure a human fisher uses from the riverbank. The lure attracts fish, which think it is something worth eating: when the lure is touched the mouth shuts in an automatic reflex. The anglerfish, which live at the deepest levels add a light to their lure: a cold light produced biologically by bacteria that have struck up a symbiotic relationship with the fish. The core group at such depths is the Ceratiidae or sea devils: a group full of picturesque names that go well with their outlandish appearance. Or should that be outoceanish appearance? The subgroups include prickly sea devil, and also black, wolftrap, needlebeard and whipnose sea devils; if ever I write a sequel to *The Silver Chair* I shall use these names for a group of wicked gnomes. These fish tend to have huge heads, vast crescent-shaped mouths with long fang-like teeth that are angled backwards. When they take hold of something they are not inclined to let it go.

There is not much food here, which is why anglerfish have resorted to the extreme tactic of the glowing lure. You won't find prey by looking for it; there's not enough of it about. You have to bring it in. This is not the teeming life of the coral reef: these are the wide – and high – open spaces of the ocean where a fish can be alone in the bleak three-dimensional wastes. And this poses yet another problem. It's hard enough to find food: how the hell are you supposed to find a mate and have sex and make more anglerfish?

The answer is as mysterious as the lure. When anglerfish were first being recovered from the deep ocean, scientists were puzzled. Why were the specimens they caught always females? And why did these females always have several parasites attached to their bodies? Where were the males? How did they have sex? The answer to all these questions was already before them. That's no parasite: that's my husband. Or one of them. The males are born stunted and minute. In some species the male's alimentary canal is so rudimentary that he can't even eat, he must find a female or die. These tiny males have one – make that two – talents. The first of these is the ability to find females in the trackless wastes of blackness. They have a high-grade olfactory system that allows them to detect the pheromones put out by females, and to do so over immense distances. They lock onto a distant female and swim towards her as if their lives depended on it, and for the very best of reasons. Once a male has found a female he bites into her skin and releases an enzyme. This digests both the skin of his own mouth and the outside of her body, eventually fusing the two together. The male locks into the female's circulatory system and obtains nutrients that way. After that, he performs his second best trick when required and produces sperm to fertilise the female's eggs.

Queerer than we can imagine. There is not a single place where it's impossible to demonstrate that eternal fact of life: not a single environment, not a single ecosystem, not a single phylum, not a single class, not a single order, not a single family, not a single genus, not a single species. If you could dive down to the places where you leave light behind you would find that, weirdly, almost horribly, you don't leave life behind. It's there in all its glorious queerness.

The wasp and the devil's chaplain

Wasps are the express route to the problem of evil. It starts with the question of why God created wasps to spoil our picnics and moves on at breakneck pace to forms of behaviour that are still stranger and more sinister. So much so that Darwin wrote: "I cannot see as plain as others do, and I should wish to do, evidence of design and beneficence on all sides of us. There seems to me too much misery in the world. I cannot persuade myself that a beneficent and omnipotent God would have designedly created the Ichneumonidae with the express intention of their feeding within the living bodies of caterpillars..."

And that's exactly what they do. Ichneumons are wasps that possess what looks like an extra-long sting. It's in fact an extra-long ovipositor or egg-laying device. The sting evolved from the ovipositor,* which is why only females can sting you. But the ichneumons have no ambition to cause humans pain, nor would it be any use to them if they did. They use the ovipositor to lay an egg inside a living creature, generally a caterpillar or a pupa. The ichneumon larva hatches out inside the creature and proceeds to eat its host from the inside while keeping it alive, which means its food will stay fresh. Stephen Jay Gould compared the process with the hanging, drawing and quartering of traitors: "As the king's executioners drew out and burned his client's entrails, so does the larva continue to eat fat bodies and digestive organs first, keeping the [victim] alive by

---

* In other words the two are homologous, like lungs and swim bladders.

preserving intact the essential heart and nervous system."

Some of the larger species of ichneumon are so adept at the task that they can lay an egg into a grub through a thick layer of wood: seeking out larvae by listening for them from the outside and then injecting the egg straight through the larva's protection system into the larva itself. There is an impressive ingenuity in all this, but it's not the sort to appeal to those who look for hummingbirds while ignoring the *Loa loa* worm. There are about 60,000 species of ichneumons, so this is a way of life that works very well indeed. Jennifer Owen spent 30 years documenting the insects in her garden, no enormous area, in suburban Leicester. In the course of her studies she identified 1,602 species of insect which included a staggering 529 species of ichneumon, of which 15 were new to Britain and four new to science. In further studies she added 74 more ichneumon species.* The ichneumon model is no one-off example of super-evil: it's a raging success, and therefore commonplace.

Wasps can be defined as all members of the order Hymenoptera and the suborder Apocrita that are neither bees nor ants. These include many species of hunting wasp, which is a descriptive rather than a taxonomic term, covering a large number of wasps that live in a similar way without necessarily being closely related. Many of these provide touching examples of maternal care. When an egg has been laid, the female provides her progeny with a larder. She collects food, mostly other arthropods. But she doesn't kill them. If she did that they'd go bad and the emerging larva would have nothing worth eating. Instead she paralyses them, and carries the inert but still living creatures, often spiders, back to the waiting egg. Gerald Durrell wrote: "If the spiders are small there may by anything up to seven or eight in a cell. Having satisfied herself that the

---

* A story told in the indispensable *Bugs Britannica* by Peter Marren and Richard Mabey.

food supply is adequate for her youngsters, the wasp then seals up the cells and flies off. Inside the grisly nursery the spiders lie in an unmoving row, in some cases for as much as seven weeks. To all intents and purposes the spiders are dead... thus they wait, so to speak, in cold storage until the eggs hatch out and the tiny grubs of the hunting wasps start browsing on their paralysed bodies."

In one of his stand-up routines, Steve Martin makes an appalling sexist joke and then tops with the irrefragable statement: "Comedy is not pretty." Nor is life. It's either funny or it's not; it's either living and procreating or it's dead, extinct and gone.

Darwin wrote again: "What a book a devil's chaplain might write on the clumsy, wasteful, blundering, low and horribly cruel works of nature!" A devil's chaplain has untold amounts of material to work on, but it would not confound any thoughtfully religious observer of nature (like, say, Simon Conway Morris), or for that matter, any serious theologian.* The existence of ichneumon wasps and hunting wasps is not proof that God doesn't exist, any more than these species' behaviour is proof that there is no such thing as, or no need for, morality. All these creatures prove is that there are no easy moral lessons to be learned from the non-human world: or to turn it the other way round, the ten million or so species on earth do not exist in order to teach humans how to operate human society.

And besides, I have a fond and treasured memory of hunting wasps. I was waiting at the entrance of the South Luangwa National Park while the formalities of paperwork were going on, when my older boy, then aged ten, pointed out with a sudden whoop of delight: "Look! Right there! Just like we saw on David Attenborough!" And there indeed were hunting wasps excavating the needle-narrow holes in

---

* A species conspicuously absent in Richard Dawkins's fundamentalist tract *The God Delusion*.

which their eggs would be laid and the "grisly nursery" established. I would never have noticed them without his keen eyes, his keen mind. In rich content we carried on to share more adventures in the bush.

The sinking fish

You wouldn't think a fish that sinks was the most efficient piece of design, but it works for the most feared predator on the planet. Sharks, like your author, are negatively buoyant. They operate from a design that is strikingly dissimilar to the ray-finned fish we have just met. The skeleton is different, as we shall see in the next chapter but one; in fact, the entire body plan is different. The swim bladder is central to the ray-finned fishes' way of life, but sharks don't have them. That's one of the reasons why my Balinese shock-shark looked so radically different from all the other fish around: it was moving forward with such great purpose. The reason sharks don't hang about in the ocean is because they can't.

Sharks need to pass plenty of water over their gill slits in order to breathe, and most species can only do this by moving. A few species, the nurse shark, for example, can rest up on the sea bottom and push water over their gill slits, but for most sharks, life is about keeping on the move: moving with crisp purpose. Sharks can swim forward while sleeping: they have to. Awake or asleep, every one of them looks like a shark on a mission, which is perhaps one of the reasons why they always look alarming. Every time you see a shark it looks as if it's after something, probably you – but usually all it's trying to do is breathe. Woody Allen's famous line in the film *Annie Hall* is based on proper natural history: "A relationship, I think, is like a shark, you know? It has to constantly move forward or it dies. What we have on our hands is a dead shark."

In this book I have constantly touched on humankind's innate inability to grasp the true nature of the diversity of the wild world. It's been suggested that we don't have the mental equipment to do so: that when we reach a certain number of different sorts of animals, our brains come up with an atavistic kind of "memory-full" response.* We can appreciate the idea that there may be more shark species than the great white, but very few of us could get into double figures. And yet there are more than 470 species, ranging in size from the dwarf lantern shark at 17 cm, 6.7 inches, to the whale shark at 12 m, nearly 40 feet. We can accept that there may be many more kinds of wasps than we know about, but we can easily write that off as mildly interesting, a matter for specialists, not really our concern. Sharks, on the other hand, fascinate humans as much as any other group of animals – and yet we still can't begin to cope with their diversity. Sharks, like practically every group on this planet, are not only more diverse than we suppose, they're also more diverse than we *can* suppose.

Of all the shark species, only four have been known to make unprovoked attacks on humans: the great white, tiger, bull and oceanic whitetip. Between 2001 and 2006 there was an average of 4.3 unprovoked attacks each year across the entire world. Knowledge of that stat won't, of course, stop me repeating my penguin impression if I have another encounter, nor will it affect the deep, almost reverent feelings of terror that sharks inspire in us humans. It seems we have a need for these monsters. And even if their human strike rate is low, it has to be admitted that their equipment for attack is singularly impressive. Their teeth operate on a sort of conveyer belt, constantly renewed. It has been

---

* See *Naming Nature*, by Carol Kaesuck Yoon, subtitled *The Clash Between Instinct and Science*. She suggests that humans can't cope with more than 600 groups of animals. She got two scientists, one her husband, to list all the genera they knew, expecting them to blow this theory out of the water. Both got as far as the high 500s and got stuck – unable to recall even animals that were familiar parts of their studies.

estimated that some sharks get through as many as 30,000 teeth in a lifetime, sometimes replacing a tooth in just eight days.

The pupils in their eyes expand and contract like those of mammals, and unlike those of the ray-finned fishes. Their sense of smell operates at long range and the distance between their nostrils gives them the ability to locate the direction of the scent. It has been suggested that the extreme adaptation of the hammerhead is designed to maximise that distance and give the most precise cross-bearing possible. All sharks hear well. They can also detect the electromagnetic fields that all living things produce; the sharks' sense of electroreception is the best in the Animal Kingdom. Sharks are considered to be intelligent and curious, and there are anecdotal reports of them indulging in play.

Not all sharks are apex predators like the great white, tiger, blue, mako and hammerhead. The dogfish* – these days more often referred to as the cat shark – is a charming little shark you sometimes find in rock pools on British beaches. The cookiecutter shark is a small shark with disproportionately enormous teeth. It has never been observed feeding, but the surmise is that it launches itself at much larger fish, seizes hold, makes a tight seal with its unusually thick lips, and then twists away violently to make off at high speed with a single mouthful of fish – leaving behind an almost perfectly circular scar, as if someone had, indeed, been using a baker's tool.

One of the weirdest sights I have ever seen in British waters – or any waters for that matter – was a group of five basking sharks, all between 10 and 12 feet, 3 and 4 m. You could estimate the size by taking the distance between the dorsal fin and tail-tip and doubling it. These are filter-feeders, sharks that cruise at their ease through the oceans

---

* They used to be served up in fish and chip shops under the euphemistic name of rock salmon.

feeding on plankton with a gaping ever-ravenous mouth. The whale shark is another filter-feeding shark, and it is bigger than all other species formerly classed as fish.

But let's get back to the great white shark. And I wouldn't like you to get the idea that it's harmless. It is a superbly effective ambush predator, up to 6 m, 20 feet, long and preying on tuna, other shark species, dolphins, porpoises, small whales and seals. The great whites are not necessarily out to get you, but I'd hate you to think that they weren't fearsome. The famous book and film exaggerated their taste for human prey, but they're still some of the finest, most brilliant and most glorious monsters on earth.

The best butter

A shard of flying light, yellow as butter, bright as if freshly churned, greeting the spring, the first to emerge that year, floaty and fluttery and as fragile as a nice day in England. A butterfly then, a brimstone, the insect that put the butter in butterfly, whose creamy-yellow colouring gave the name to thousands of species across the world. "It was the best butter," as the March Hare said meekly to the Mad Hatter, having used it to fix the Hatter's watch.[*]

So now let's switch to a summer evening, a family gathering round the table, much laughter in the warm air. It had been dark for an hour or two now, and what's that fluttering round the candles seeking the flame with such dreadful certainty? I gathered the moth in my cupped hands, a hawk moth, so large it was almost an honorary bird, a poplar hawk moth, I think it was, but I didn't stop to examine it for long, and as it tickled my palms in all its eternal vulnerability, I took it some distance from the light and lobbed it skywards, willing it to find a better target for its passion.

Butterflies and moths. Lepidoptera, which means scaly wings. How many species? One estimate gives me an alarmingly precise 172,250 species, There are rough and ready differences between butterflies and moths, but it's not a true taxonomic division: better accept that there are no moths and no butterflies, much as there are no fish. Plenty of moths make their living in sunlight: the six-spot burnet is rather

---

[*] From *Alice in Wonderland*, of course.

a favourite of mine, and the silver Y moth is the butterfly you can't identify as it sups nectar from the flowers. I have a neighbour, David Wilson, who did the illustrations for *Colour Identification Guide to Moths of the British Isles*, and he did a rare and splendid thing for me. I had spent a fine evening with him, in which we raided his moth traps, drank a beer, raided and drank again and so forth: an excellent way of doing natural history. I am no great shakes at moth identification but it is always good to spend time with people prepared to explain the passions of their lives. And I told him I had always wanted to see a hummingbird hawk moth.

A few weeks later he rang me up to say that one was currently visiting his garden if I could get along quickly. Alas, I was covering a cricket match many miles away and couldn't make it. After a few more weeks he invited me round again. He had caught the moth, and when it had laid eggs, he had released it. He had hatched the eggs out and fed the caterpillars and he now had a fully emerged moth. This time I was able to get round right away. At his invitation, I released the moth, which crawled onto my finger and savoured the air of the Suffolk garden before committing itself. We joked that it would probably die on the spot or get eaten by a hungry bird straight away, but not a bit of it. After a while it took to the air and flew straight to the adjacent buddleia bush and started to take nectar: hanging in the air in exactly the manner of a hummingbird. Every year many people ring conservation organisations to report that they have a hummingbird in the garden, but it's a British moth pulling off exactly the same trick. It's yet another example of evolutionary convergence.

If the ichneumons are the insects for the devil's chaplain, then butterflies redress the balance for the opposition. To see a butterfly is to be convinced that the world exists for nothing but peace, love and beauty. I recall a butterfly in the Luangwa Valley, right at the end of the dry season in the ferocious murdering heat, as lovely a thing as I have seen: black, painted in iridescent green streaks, with two long and elaborate tail streamers. This, I eventually worked out, was a cream-striped swordtail, each wing the size of my palm. Back in England there is a butterfly nearly as extravagant, one you can find in season in the more obscure parts of the Norfolk Broads. I found swallowtail for the first time at Hickling: improbably large and improbably bright and yellow-patterned, moving with a sense of urgency over the tops of the reedbeds.

Butterflies are unambiguously lovely: moths in their muted way not much less so, even though they are associated with such sinister things as night and darkness. It is not necessary to say you find butterflies attractive: it's an instinctive part of human nature. Yet we find caterpillars mildly repulsive. I remember a cartoon of two caterpillars looking at a butterfly: "You'll never get me up in one of those." Butterflies, like many insects, compartmentalise their lives: as if we humans spent the first 65 years of our lives doing nothing but eat, and then gave up solid food and, getting a bus pass, spent our remaining few months or years as full-time sex maniacs.

Every now and then, Britain gets a painted lady year: a year with vast numbers of this rather special species of butterfly. They migrate up from Morocco in wildly fluctuating numbers. For a long time this migration was considered an evolutionary anomaly, or Darwinian cul-de-sac. The butterflies flew up to Britain, arrived in numbers, bred: and then simply vanished. It was assumed that this was a doomed generation, an experiment that continually went wrong, the butterflies pushing too far north to make a go of it. But in recent years painted ladies have been observed making the return journey: a greater miracle even than their miraculous appearance.

In Britain it is generally accepted that there are 59 species of butterfly and around 2,000 of moths, including the tiny and mysterious micromoths. In the world there are, as we have seen, not far short of 200,000 species of Lepidoptera. Many of them are staggeringly, outrageously, almost absurdly beautiful. And how many names do we have for them in common use? Two. Butterfly and moth. You could hardly get better evidence for the inability of non-specialised human vocabulary – and therefore of humankind – to come to terms with the diversity of the Animal Kingdom.

So let me close this chapter by telling you the easiest way of increasing your own appreciation of biodiversity. You don't even have to leave your own garden. Most of us can name two or three butterfly species. Get a book* and see if you can't get into double figures. And that's it. Simplicity itself. You can do it with a cold drink in your hand: sip, look, and then look for a name. Part of the magic of names is that they make you look more closely and more often. A name changes the way you look *for*, and once known, a name changes the way you look *at*. Sitting in a garden on a nice sunny day, preferably with a nice sunny drink, is an exercise in increasing your understanding of biodiversity: that is to say, understanding the basic mechanism that supports life on this planet.

---

* Try *A Pocket Guide to the Butterflies of Great Britain and Ireland* by Richard Lewington.

No bones about it

It seems obvious now that the taxonomists have explained the puzzle. How could sharks and their relatives ever be confused with the ray-fins? The difference goes bone-deep. I'd say marrow-deep, but sharks don't have bone marrow, not having bones as we know them.* Instead, they have a skeleton made of cartilage: flexible, tough, lightweight, elastic. They start off with a notochord, the incipient spine that is found in all vertebrate embryos, our own, of course, included, but with the cartilaginous fish this is gradually replaced by cartilage. It is a unique strategy among us chordates, which explains why the group – the Chondrichthyes – have so singular an appearance.

The weirdest are the chimeras or ghost fish, 34 species and all of them odd, even monstrous. They are sometimes called rabbitfish, because of their fused, plate-like teeth, and they include ratfish, elephant fish and charmingly, a spookfish. They look rather like bad drawings of fictional monsterfish intended to horrify a not overcritical audience of children. There are some creatures that, when you see them illustrated, or even photographed, look so strange you can't believe they seriously exist. They look too much like an artist's fantasy.† The chimeras demonstrate that principle to perfection.

The second and much larger group, the Elasmobranchii,

---

* So how do they make red blood cells? The answer is the spleen, the epigonal organ which is found near the gonads, and the specialised Leydig's organ. The chimeras possess neither of the last two.
† Among us mammals, the aardvark is the leading example of this principle.

includes sharks, already discussed, and rays. Sharks and rays have taken opposing strategies. Sharks, as we know, are streamlined and forever moving forward. Rays and skates have gone the other way, abandoning streamlining and sinking to the bottom. Sharks must keep moving forward in order to breathe, for they need the passage of water over their gill slits. The rays have taken a different evolutionary direction and stillness is no problem to them: in fact, it is the foundation of their existence. They breathe by taking in water through their spiracles, small holes behind their eyes, and out through their gills. They are adept at concealment, and feed on other bottom-dwellers, mainly molluscs and crustaceans.

They have taken on a flattened, non-streamlined shape, with their pectoral fins fused to their heads. They swim with a rippling motion, for preference close to the sea bottom, so that they seem to glide frictionlessly over the floor, like the puck in a game of air hockey. Their eyes are on the top of their heads, which is fine for spotting potential predators, but it means they can't see the animals they pursue themselves. They hunt them out with smell and with electrical receptors instead: an extreme example of compartmentalisation of the senses. They have a reputation for intelligence, some say they are smarter than sharks.

Stillness makes an animal vulnerable, so the rays have evolved more than one counter-strategy. They bury or half-bury themselves in sand, and most species are neutral coloured and tend to blend into the background. Some species are well camouflaged. Stillness can be effective, since it's movement that catched the eyes, on solid ground or underwater, but all the same it helps to have a plan B. Anyone who has taken a walk in the English countryside knows about plan B: a pheasant will keep dead still until you are right on top of it, and then it will leap clumsily into the air while making the most terrible din. It makes you jump: it's supposed to. The noise is designed to freeze the predator for a vital half-second while the pheasant[*] escapes. Some rays have a different but equally effective plan B: attack.[†] Stingrays carry one or more spines on their tails equipped with venom, and will use it in self-defence. Some of the big species carry truly fearsome weapons, with a sting 35 cm or 14 inches long.[‡] Steve Irwin, the Australian television presenter, died after being pierced in the chest by a stingray.

The group also includes the sawfish – and even as I write the words I seem to recall a sawfish displayed in Birmingham Museum not far from the Japanese spider crab. It was worth a good stare: a toothed blade that looked infinitely capable of sawing down an oak. It is used as a kind of scatter-gun weapon for slashing through a shoal of small fish, and it has a secondary use as a mud probe. Sawfish can reach 6 m, 20 feet, in length, with a saw 6 feet, 1.8 m, long and 30 cm or a foot wide. You look for an evolutionary pattern, find it – and then just about every time, you come across an oddball, something that really doesn't fit in. Moral: wildlife is not trying to please the human mind. It seldom conforms to the human mania for tidiness.

---

* That's why pheasant shooting is an unfair competition: the pheasant strategy is to lie low, and then as a last resort, to make a predator jump. So the beaters startle the pheasant into flight – and only then, when the pheasant has exhausted its evolved options and is vulnerable, do the guns open fire.

† A classic example of the old adage: "This animal is dangerous: it defends itself when attacked."

‡ There are stories of a stingray's tail being used by humans as a whip to chastise slaves and errant wives. The British government made them illegal in Aden. A stingray whip appears in the James Bond short story, already quoted in his book, "The Hildebrand Rarity".

Inordinate fondness and all that

Beetles have four wings, in the classic insect pattern, but their forewing has become a hard, thick sheath. Beetles make up the order of Coleoptera, which means sheathed wing. When they fly – most beetles fly – they flip up the modified forewings and leave them there, so they look as if they are flying with the doors open. In this fashion they power themselves along effectively enough with their hindwings. Which means they can move in confined spaces without damaging their wings, so they can still fly.

What's so bloody marvellous about that?

I mean, it's quite a neat trick, I wouldn't deny that for an instant, but it's hardly the most brilliant piece of adaptation in the entire history of life, is it? Compared to a shark's teeth, an albatross's wing, an elephant's trunk, a human's brain, a mosquito's mouthparts, a waterbear's unkillability or a termite's social life, the fact of sheathing a wing doesn't strike me as one of the Beamonesque* evolutionary leaps in the story of life on this planet. No one making a display of the wonders of nature for children would say that the most wonderful thing of all is that beetles can tuck their wings away behind a small fragment of armour. Sheathed wings are not going to inspire gasps of amazement and

---

* Bob Beamon (already met in these pages as a comparison with leaping bushbabies) was the athlete at the Mexico Olympic Games who broke the world long-jump record of 29 ft 2½ in, 8.90 m, beating the previous best by an impossible 21¾ in, 55 cm. He set a mark that wasn't beaten for almost 23 years.

a determination to look after the non-human life of the planet better than before. But this adaptation has made beetles the most impossibly, the most brilliantly, the most startlingly diverse aspect of life on earth. Of life in the entire universe, for all we know.

Let's take some numbers: 25 per cent of all known life forms are beetles; 30 per cent of all animal species are beetles; 40 per cent of all described insects are beetles. There are getting on for 400,000 species of beetles already described and more are being found every year, or every day. Estimations for the number left to be described vary from a million to 100 million – and even a mere million is quite a lot, if you think of that in terms of pounds or dollars. The most famous line about beetles comes from JBS Haldane, who we have already met in these pages via his queerer-than-we-can-suppose remark about the universe. The story – there are many variants – is that Haldane was asked by a theologian what he could deduce about the nature of the Creator from a lifelong study of His Creation. Haldane's response: "An inordinate fondness for beetles."* In an essay written later, Haldane expanded the thought: "The Creator would appear as endowed with a passion for stars, on the one hand, and for beetles on the other, for the simple reason that there are nearly 300,000 species of beetle known, and perhaps more, as compared with somewhat less than 9,000 species of birds and a little over 10,000 species of mammals. Beetles are actually more numerous than the species of any other insect order. That kind of thing is characteristic of nature." (Note, by the way, that Haldane's numbers are no longer in fashion.)

Alfred Russel Wallace, Darwin's co-discoverer of the

---

* I have two books of this title; the first a textbook about beetles by Arthur V Evans, and the second a splendidly flaky account of Alfred Russel Wallace by Paul Spencer Sochaczewski. The quotation by Wallace in this chapter comes from this book.

principle of evolution by means of natural selection, once wrote in his notebook: "It is a melancholy thought that many of our fellow-creatures do not know what is a *beetle*! They think cockroaches* are beetles! Tell them that beetles are more numerous, more varied and even more beautiful than the birds or beasts or fishes that inhabit the earth and they will hardly believe you, – tell them that he who does not know something about beetles misses a never failing source of pleasure and occupation and is ignorant of one of the most important groups of animals inhabiting the earth and they will think you are joking, – tell them that he who has never observed and studied beetles passes over more wonders in every field and every copse than the ordinary traveller sees who goes round the world and they will perhaps consider you crazy, – yet you will have told them only the truth."

Darwin spent much of his youth in pursuit of rare beetles, much as modern twitchers chase rare birds across Britain, filled with love of wild things and inflamed with collection mania. (Perhaps if Darwin had had a car and a mobile phone and a pair of top-of-the-range modern binoculars he would have become a twitcher, for it is much the same sort of obsession.) Darwin wrote in a letter: "But no pursuit at Cambridge was followed with nearly so much eagerness or gave me so much pleasure as collecting beetles. It was the mere passion for collecting, for I did not dissect them and rarely compared their external characters with published descriptions, but got them named anyhow. I will give a proof of my zeal: one day, on tearing off some old bark, I saw two rare beetles and seized one in each hand; then I saw a third and new kind, which I could not bear to lose, so that I popped the one which I held in my right hand into my mouth. Alas it ejected some intensely acrid fluid, which

---

* Cockroaches make up an order of their own, the Blattodea. There are about 4,500 species – about as many as they are mammals – and 30 of them are associated with humans.

◦ 14. RHINOCEROS BEETLE ◦

burnt my tongue so that I was forced to spit the beetle out, which was lost, as well as the third one."[*]

In other words, the two men who were responsible for unravelling the greatest mystery of life both suffered from beetlemania. This is not the weirdest of all possible coincidences; beetling and bug-hunting were more main-stream activities in Victorian times. During his beetling days, Darwin thought he might fancy the life of a parson naturalist, and would have joined a long tradition had he done so. It is tempting to describe beetle numbers as "head-spinning" or "mind-boggling", and so they are for most of us. But neither Darwin nor Wallace had minds that did much boggling. When their minds were stimulated they looked for solutions. Darwin's mind always reminds me of a description of Karz, the bad guy in one of the Modesty Blaise novels.[†] "There was finality in Karz's voice. The sloe-black eyes went suddenly blank. Liebmann knew that Karz had finished with the subject, withdrawn his mind from it. His thoughts were occupied by some other aspect of the massive and complex operation. He might sit for five minutes or five hours, until his mountainous mind had crushed the problem out of existence." Karz essentially represents the forces of anti-life: Darwin quite the opposite. Darwin's mind was mountainous all right, but his problem-solving nature was profoundly life-affirming. There is a deep joy to be found in Darwin's works, and it comes first from the writer himself. That joy can be found again in the way we look at life on earth after he opened our eyes to its meaning.

And the fact is that Darwin did crush the problem, slowly and laboriously and inevitably grinding it to a pulp with the rock-crushing mechanism of his mind. It is easy to tell what creatures took him there: everything he had ever seen and

[*] This was the crucifix ground beetle, thought extinct in this country but recently found in Wicken Fen in Cambridgeshire.
[†] *Sabretooth*, quoted earlier with reference to whelks.

chased and shot and caught and observed and missed and spat out. But my view is that it was the inordinate numbers of beetles, the extraordinary and apparently endless variety of beetles, that first told him – in a subtle, sleeping, unrealised form – that there was a massive question out there that needed answering.

As for the beauties of beetles, come boating with me in Southeast Asia. I have experienced this miracle twice, in Borneo and in peninsular Malaysia: a riverbank illuminated by beetles: an entire tree that turned itself on and off, like a neon sign in Piccadilly Circus or Times Square, lighting up the branches and the quiet waters beneath with a cold and mysterious otherworldly light, like the lanterns carried by the gnomes of CS Lewis's Underworld. In stately, deliberate rhythm they flashed on, they flashed off, they flashed on again. Here was a group, a gathering, a swarm of fireflies all acting as one, and for the same rich purpose: to find mates, to make more fireflies. Americans call them lightning bugs; there is a species that goes in for the same sort of synchronous flashing in Great Smoky Mountain National Park. And these flies are not flies, and these bugs are not bugs. They're all beetles, like practically everything else that lives: part of the overwhelming, world-solving, inordinate numbers of beetles. So I should add, then, that there are 2,000 species of fireflies or lightning bugs, and there are some that will flash the code of the wrong species to dupe an ardent mate-seeking firefly and then devour it. Adaptations within adaptations within adaptations: queerer than we can suppose: just a couple of the beetles' greatest hits.

Ray of sunshine

Who has not played the great game of counter-factual auto-biography? I suspect all three of my published novels started from that premise, the third from the almost universal experience of wondering what would have happened if instead of breaking up so traumatically, you and that person from back then had got still more traumatically married. What if I'd gone to Africa instead of Asia when I was young? What if I'd stayed in Asia instead of going back to England? What if I'd gone to Australia? What if I'd discovered horses when I was young enough to have got seriously good? What if I'd learned to dive? What if, negatively buoyant, I had taken to the sea and fallen in love with cetaceans and all those wonderful things we sometimes call fish, instead of birds and the big mammals of the savannah?

I was having dinner with a family, old friends all, and in the course of it I let slip that I was planning a chapter on manta rays. As one person, they chorused: how wonderful these creatures are, how seeing them, swimming with them, being with them was one of the most wonderful experiences of all their lives. For they all dive, often all four together, and are all well-practised diving buddies. They have been doing this since the younger ones were children. But I have never dived, never seen a manta ray, and my life is poorer as a result. My imagination has always been haunted by these most unearthly, most unoceanly of creatures: creatures from another world, oceanic angels with broad, broad wings, creatures twice as wide as they are long, paying a brief visit to this strange planet full of aliens ten million and more strong. I can see myself beneath as one passes, turning my entire body, not just my neck, as you can when moving almost weightless in three dimensions, to stare *up* at the great shape above: the vast winged cloud interposing itself between me and the sun, not swimming, not even flying, but *gliding* overhead.

Manta rays seem to me creatures with something of the miracle about them – but then that's true of everything that walks or crawls or swims or flies or wriggles or squirms or anchors itself to the ocean bed and filter-feeds. It's just that these great lovely flying monsters seem to me to express that touch of the miraculous better than most. Perhaps they are all the more vivid because I have only seen them on a television screen and of course, in my imagination. They are rays, and rays, as we have seen, evolved for the floor of the ocean, but the manta rays – there are reckoned to be two species – left the bottom and took to the open seas once again and did so with grace and magnificence.

They are filter-feeders, feeding on zooplankton, helping the stuff into their gawping mouths with the help of two flappy fins on their heads – cephalic fins, technically. They can measure getting on for 7 m, 23 feet, across, and astonishingly, they sometimes breach, that is to say, jump clear of the water. They have even been seen performing somersaults. Like sharks they need constant movement if they are to stay alive: the passage of water over their gills sustains their lives. They are long-lived – some individuals are reckoned to be at least 50 years old. They sometimes gather in groups of 50 and more, and they form associations with cleaner fish; often other fish come along for the ride to pick up falling food items. They are warm-water specialists, preferring tropical and subtropical latitudes. They often come close to the surface in shallow water, especially in coastal waters. In other words, they are a pretty accessible miracle for those with a taste for diving.

It is their mountainous, seemingly magical presence that

piques my imagination so strongly, but there is another point that needs consideration. The ancestors of these great exponents of perpetual movement and undersea flight evolved for the bottom and for stillness, abandoning the life of the constant swimmer. And that really doesn't make logical sense. It seems like shilly-shallying: a lineage, a set of genes, set off to do one job and then changed its mind and went back to the original idea.

This open-water, sea-bottom, open-water switching back and forth is confusing, but it's an evolutionary lesson and an important one. Mammals are ultimately descended from backboned creatures that lived in the sea and went onto the land in search of a living. Many mammals have returned to the water: some part-timers, like otters, some most-of-the-timers, like seals, and some all-the-timers, like whales and dolphins. Where is the progress in that? If you've got a plan, stick to it: don't change your mind halfway. But these rules of human planning don't operate in the wider context of life. Evolution is not a plan, and it's not about progress. Some cartilaginous fish found bottom-dwelling an effective way of life and specialised: some rays found that moving back into open water worked pretty well. The aim of life is not to live better than your ancestors: it is to do anything that helps you become an ancestor yourself. That is the secret behind every chapter in this book: and in every one of the ten million or so other aliens that I somehow missed out along the way.

Axis of weevil

So, Mr Barnes – Simon – what was your favourite? After all those chapters, all those words and all those creatures, what's the best animal in the Animal Kingdom? What's the most effective, the most formidable, the most lovely? What is the champion of champions: the one creature that stands out, not just from the phylum of vertebrates but from all the multiform unrelated phyla of invertebrates as well? Where is it that the Animal Kingdom reaches its peak? What is the masterpiece of Animalia? What is the animal of animals?

I have spent most of this book trying to explain that life's not like that, that life doesn't work in this way, that life's not about champions and linear progression and hierarchies. And it really isn't. But our human minds don't accept the world as it is. Our minds don't have the appropriate functions. We'd much rather have the King of the Jungle, with everything else descending from this summit in an orderly fashion, getting feebler and more insignificant as we get closer to the bottom. Especially when we take the role of King of the Jungle for ourselves.

But this book is a circle, or rather two circles – and above all it's about diversity. If I'm going to gratify our human urge for a champion I'd better find something that fits the theme of this book. So let's settle for the weevil. There really is nothing like a weevil. Or rather, there is: there's an awful lot of things so exactly like a weevil that they're weevils too. If there is a champion at all, it is to be found not in conventional magnificence or in power to fire the human imagination, but in multiplicity, in diversity, in the ability to boggle, in the ability to make us understand for all time that life on earth really is queerer than we can suppose.

Weevils are beetles. In fact, it's probably true to say that what insects are to the rest of the Animal Kingdom and what beetles are to insects, so weevils are to beetles. In diversity there is nothing to touch them. Weevils make up one single family. That's a low-level category of classification: species, genus, family. Let's not make comparisons with human classification – that will stir up a load of controversies that we've already dealt with and which won't help us here. So let us take cats instead: after all, it's the natural place to look for favourites and champions and kings of the jungle. Let's take the family Felidae. It ranges from tigers to next-door's cat: it includes lynx, puma, ocelot, caracal, serval, snow leopard, jaguarundi, fishing cat, the wildcat of Scotland and elsewhere and the delightfully named flat-headed cat, an elegant beast I once saw in Borneo. A large and diverse family, then: one that contains about 41 species. The family of weevils, the Curculionidae, has 40,000 species. A thousand times greater: greater by three orders of magnitude.

Snout beetles and bark beetles. They have a long nose – technically a rostrum, as we've already seen – but it's not a poker and piercer, in the manner of the true bugs'. The weevil rostrum has tiny chewing parts right at the end of it. As with sheathed wings, it's tempting to ask what's so bloody marvellous about that, but it's the device that has set loose this bewildering adaptive radiation of weevils. They are none of them terribly big: the size range is 1–40 mm, 0.04–1.6 inches. They are mostly vegetarians; in fact, there is scarcely a species of plant on the planet that doesn't attract weevils. Weevils are capable of exploiting just about every bit of plant that grows: wood, roots, leaves, seeds, fruit, flowers, shoots.

Weevils represent one of the greatest miracles of life on earth: which is to say the twin miracles – Siamese or conjoined twins – of diversity and adaptability. Of course, if you look them up on the internet, you'll find very little about them except how to kill the ones that get in the way of human life. We ignore weevils apart from when we want to kill them, yet weevils tell us, more clearly and more vividly than any other family of animals on this planet, about life: what life is about, how life works and how extraordinary life is. Should I push this line of thought still further and say that weevils also tell us how beautiful life is? Weevils are not beautiful to human eyes, not like tigers and flat-headed cats: but they are at least beautiful in conception. We should look on weevils as we look on cubist paintings: admiring the thought, the execution, the all-round multidimensional perfection – even if we'd rather have one of the impressionists in the previous room of the gallery hanging on our bedroom wall. The cubists changed the way we see the world: for Braque and Picasso I give you weevils.

The beginning

One last dizzying plunge remains: a swallow-dive down to the very depths of vertebrate life – and then up and back out the other side again. We will meet species incomparably and incomprehensibly ancient as we go. There are two classes of vertebrates left, both formerly lumped together in the group called "fish", and they include some of the weirdest creatures on the planet – not least because they include, stretching a point just a little, us.

But first the jawless fish. Their lineage stretches back to the Cambrian era, to the time when life first exploded in a great detonation of diversity 530 million years ago. To understand life on earth you must first understand Time as it really is: not time with its seconds ticking their relentless way around the clock-face, but Time as something deeper than did ever plummet human mind. The only problem with this is that you can't. It's impossible for us humans to imagine a thousand years. A single million is as far out of reach as the Horsehead Nebula, and that's still part of the galaxy we live in. Half a billion years is so far beyond human scope that it might as well be fantasy. Even *Finnegans Wake* aimed to tell no more than the entire history of human civilisation – but all the same, reading that extraordinary book is perhaps the nearest we can get to grasping the impossible immensities of Time. Deep Time is the essential ingredient for shaping the earth and its creatures into the forms they take today, throughout all the convolutions and convulsions of history, a history that included many extinction episodes, five of them drastic

– six if you count the extinction crisis of modern times.

Back in the Cambrian we find the jawless fish, not unrecognisably different from the 100-odd species that survive today, lampreys and hagfish. And no, they really don't have any jaws: just suckers and a snout and concentric circles of rasping teeth. They have a jawless cranium made of cartilage, a partially formed vertebral column and no scales. They tend to feed parasitically on other – to use the term loosely – fish.

And now to the last class of all: the last class in the phylum of vertebrates: the last class we shall meet in this book: the lobe-finned fishes. The most famous of them is the coelacanth, which was known only from fossils 65 million years old when a living animal was caught in 1938. The group also includes lungfish, those mysterious creatures that can bury themselves in mud and survive for months until the rains come again; creatures found in Africa, South America and Australia.

Lobe-finned fishes are distinctly odd because their front and back fins – their pectoral and pelvic fins – look rather like primitive limbs. They can use these to shuffle along the sea bottom. The class includes the Tetrapodomorpha, sometimes called fishapods. These look really quite a lot like fish with legs: and they can use these almost-legs to leave the water for brief periods. Here, then, represented by the last species of the book, the Tiktaalik, is the four-limbed body plan: here, then, is the basic idea of us. Here, then, is the scheme adopted by frogs and toads and lizards and crocodiles and birds and cats and dogs and mice and rats and you and me.

At some stage across the profundities of Deep Time it became advantageous to leave the water behind. There are various suggestions about what prompted this seismic change. There is the stinking waterhole hypothesis, sometimes called the desert hypothesis: the place where the ancient lobe-fins lived dried out or otherwise failed; those that could scrambled out to look for new water. Then there is the intertidal hypothesis: the twice-daily emergency of the low tides is also an opportunity for underwater creatures that can exploit the exposed areas. The mudskippers – ray-finned fishes and nothing to do with the bunch here – exploit this niche today, often among mangrove swamps, and a very fetching sight they are, too. And then there is the woodland hypothesis, suggesting that the urge to leave the water came from the advantages gained in flooded forests.

Whichever way it was, this most obscure branch of the phylum of vertebrates is the key to all the backboned creatures that live on the surface of the earth or fly above it, and quite a lot of those that live in the water as well. Perhaps we owe our existence, not to some vast eternal plan, but to the stink of a dying waterhole. Counter-factual history: what if this primitive little thing, this fish that is not a fish, this small and undistinguished group, failed to make it? It's the question I asked earlier, when we contemplated the ancestor shared by humans and xenoturbellids: for the same curious circumstance arises again and again. So: what if one of the early extinction crises caused the lobe-finned fishes to go extinct? It wouldn't register as a catastrophe when the Martians or the Tralfamadorians arrived to study the history of life on earth – and yet life on earth would have been utterly different, and we tetraprods, we mammals, we apes, we humans would not exist.

And no, life doesn't divide itself into a series of compartments and boxes. Life is moving and fluid. Its boundaries merge perplexingly one into another like the characters in *Finnegans Wake*, while the way we try and make sense of it all is changing and changing again, to the bewilderment even of specialists. The separation of humans from the rest of the Animal Kingdom no longer looks like a hard and fast thing: it's all a matter of definition and degree. And the thrilling thing is that the dissolving of the human–animal boundary does nothing but enrich us. It makes us more at home on our own planet and more capable of understanding who we are and what the hell we are doing. If there is a moral separateness of humans from other animals it is that we have a better idea of the long-term consequences of our actions: not that the Tralfamadorian visitor would deduce that from the way we actually carry on.

We are part of a continuum. We are linked to our fellow animals, linked by our past and our present, linked by evolution and by ecology. We are linked, above all, by the planet that has supported us all for so many years. So far, anyway.

And this sense of continuity is what matters: the continuity of our wild places, the continuity of the species with whom we share the planet, the continuity of the life-support system that is planet earth. We're all in the same boat: we all come from the same stock, we're all part of the same continuing cycle of life on earth. How was it Darwin ended his big book? Ah yes. There is grandeur in this view of life, with its several powers, having been originally breathed into a few forms or into one; and that, while this planet has gone cycling on according to the fixed law of gravity, from so simple a beginning

15. COELACANTH

Acknowledgements

Grateful thanks to all the following:

John Burton, of the World Land Trust, for reading the manuscript and also for providing many of the experiences that crop up in its pages

Chris Breen of Wildlife Worldwide for many shared adventures

The late Baron Robert Stjernstedt, ornithologist. Knowing him was one of the great adventures of my life

The late Aaron Mushindu of Livingstone Museum

All at Zambian Ornithological Society, especially the late Dylan Aspinwall

All at Norman Carr Safaris, including the late Norman Carr, Adrian and Christina Carr and Abraham Banda

Tim Dodman of WWF and Lochinvar National Park

The British charity Riders for Health

South Luangwa Conservation Society

Wildlife Trust for India, especially Vivek Menon and Sandeep Tiwari

The late Peter O'Donnell, author of the Modesty Blaise books

Lee Durrell of the Durrell Wildlife Conservation Trust, for permission to quote Gerald Durrell at length

All at Suffolk Wildlife Trust, especially Julian Roughton, Simone Bullion, Alison Looser and Dorothy Casey

All at Save the Rhino, especially Cathy Dean

Yorkshire Wildlife Trust and Anthony Hurd

Janie Ray of Cetacealab in British Columbia, and Neekas the whale-dog

Phillip Clapham of Alaska Fisheries Science Center

John and Carol Coppinger of Remote Africa Safaris

Nick Aslin of Zambia Ground Handlers

Diego Calderon-Franco of Colombia Birding

Nicholas and Raquel Locke of Regua, Brazil

Butterfly Conservation, especially Martin Warren

Professor Nicky Clayton of Cambridge University

All at Seawatch Foundation

Carl Chapman of Wildlife Tours and Education

Fergus and Di Flynn of Lechwe Lodge

All at El Almejal, Colombia

Ralph Bousfield of Jack's Camp, Botswana

Phil Berry, expert on Zambian Wildlife

David Wilson, moth expert

John-Paul Davidson

Stephen Fry

Emma Craigie, for her wise editing

All at Short Books, especially Rebecca Nicolson and Aurea Carpenter

Georgina Capel, agent and comforter

My family Cindy, Joseph and Eddie

Index

Page references in *italics* indicate an
illustration. 'n' indicates a reference
to a footnote.